Disclaimer

The publisher of this book is by no way associated with the National Institute of Standards and Technology (NIST). The NIST did not publish this book. It was published by 50 page publications under the public domain license.

50 Page Publications.

Book Title: Disaster Resilience: A Guide to the Literature

Book Author: Stanley W. Gilbert

Book Abstract: Although there is a great deal of high-quality information available on resilience-related topics hazard assessment, vulnerability assessment, risk assessment, risk management, and loss estimation as well as disaster resilience itself, there is no central source of data and tools to which the owners and managers of constructed facilities, community planners, policy makers, and other decision makers can turn for help in defining and measuring the resilience of their structures and communities. The purpose of this document is to provide a survey of the literature and an annotated bibliography of printed and electronic resources that serves as that central source of data and tools to help readers develop methodologies for defining and measuring the disaster resilience of their structures and communities. The report covers resilience-related topics at two different levels: (1) individual constructed facilities and correlated collections of constructed facilities that form a network (e.g., hospitals) and (2) community/regional scale frameworks (e.g., physical infrastructure, business and economic relationships, population and employment demographics). Thus, the first level focuses on physical infrastructure, whereas the second takes a broader look at how the physical infrastructure interacts with other activities that collectively define modern communities. The reason for taking this approach is to establish a foundation for developing methodologies for defining and measuring the disaster resilience of structures. This step is especially important because physical infrastructure enables the community to function as a place of employment, a window to the regional and national economy, and a home for individuals. Developing better metrics and tools for defining and measuring the resilience of structures is an important step in meeting the challenge of measuring disaster resilience at the community scale.

Citation: NIST SP - 1117

Keyword: Buildings; communities; constructed facilities; disasters; economic analysis; insurance; natural and man-made hazards; resilience; risk assessment; vulnerability assessment

NIST Special Publication 1117

U.S. Department of Commerce
National Institute of Standards and Technology

Office of Applied Economics
Building and Fire Research Laboratory
Gaithersburg, Maryland 20899

Disaster Resilience: A Guide to the Literature

Stanley W. Gilbert

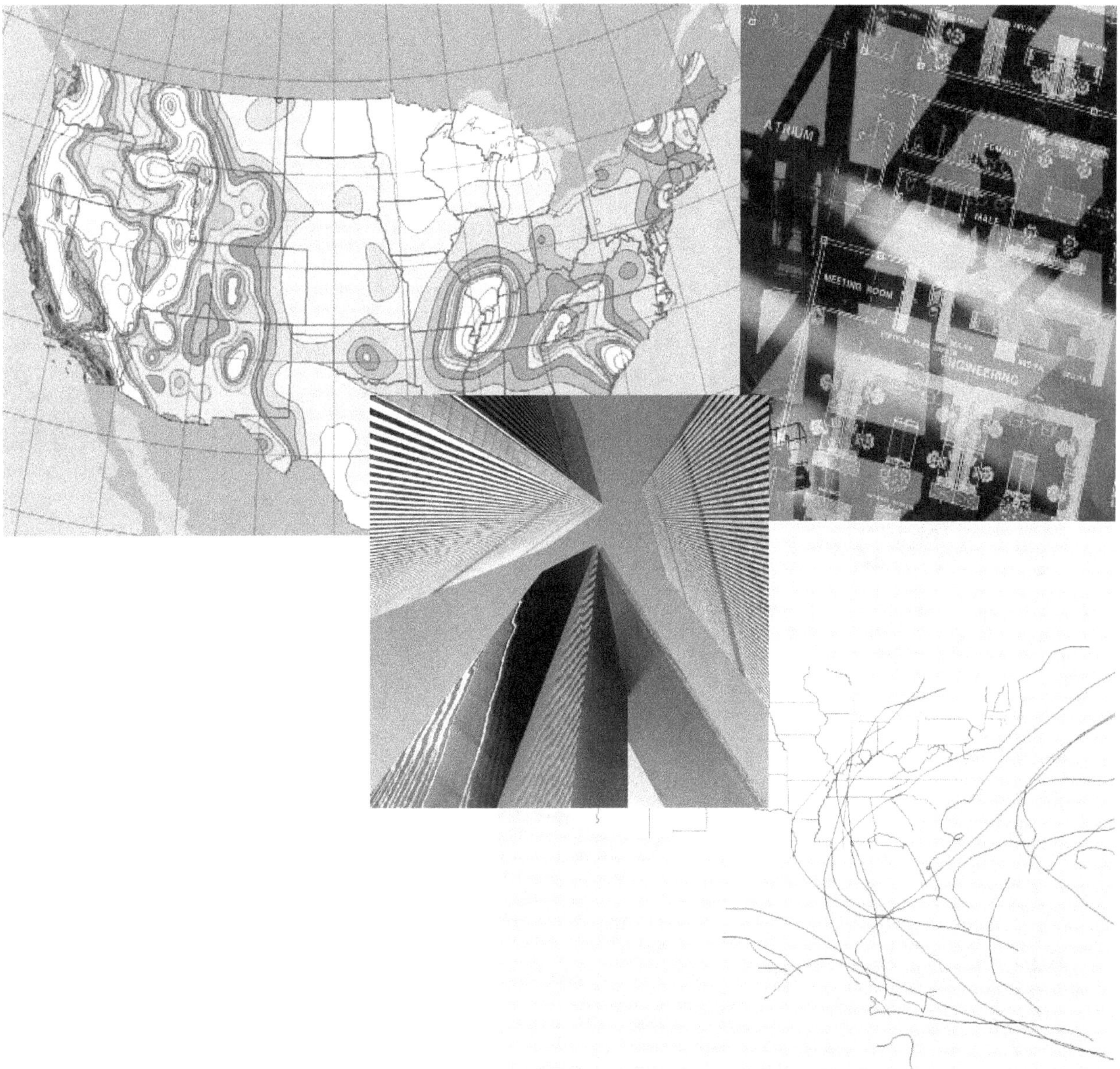

NIST Special Publication 1117

 U.S. Department of Commerce
National Institute of Standards and Technology

Office of Applied Economics
Building and Fire Research Laboratory
Gaithersburg, Maryland 20899

Disaster Resilience: A Guide to the Literature

Stanley W. Gilbert

Sponsored by:

National Institute of Standards and Technology
Building and Fire Research Laboratory

September 2010

U.S. DEPARTMENT OF COMMERCE

Gary Locke, Secretary

NATIONAL INSTITUTE OF STANDARDS AND TECHNOLOGY

Patrick D. Gallagher, Director

Abstract

Although there is a great deal of high-quality information available on resilience-related topics—hazard assessment, vulnerability assessment, risk assessment, risk management, and loss estimation—as well as disaster resilience itself, there is no central source of data and tools to which the owners and managers of constructed facilities, community planners, policy makers, and other decision makers can turn for help in defining and measuring the resilience of their structures and communities. The purpose of this document is to provide a survey of the literature and an annotated bibliography of printed and electronic resources that serves as that central source of data and tools to help readers develop methodologies for defining and measuring the disaster resilience of their structures and communities.

The report covers resilience-related topics at two different levels: (1) individual constructed facilities and correlated collections of constructed facilities that form a network (e.g., hospitals) and (2) community/regional scale frameworks (e.g., physical infrastructure, business and economic relationships, population and employment demographics). Thus, the first level focuses on physical infrastructure, whereas the second takes a broader look at how the physical infrastructure interacts with other activities that collectively define modern communities. The reason for taking this approach is to establish a foundation for developing methodologies for defining and measuring the disaster resilience of structures. This step is especially important because physical infrastructure enables the community to function as a place of employment, a window to the regional and national economy, and a home for individuals. Developing better metrics and tools for defining and measuring the resilience of structures is an important step in meeting the challenge of measuring disaster resilience at the community scale.

Keywords

Buildings; communities; constructed facilities; disasters; economic analysis; insurance; natural and man-made hazards; resilience; risk assessment; vulnerability assessment

Preface

This study was conducted by the Office of Applied Economics in the Building and Fire Research Laboratory at the National Institute of Standards and Technology. The study provides a survey of the literature and an annotated bibliography of printed and electronic resources that serves as a central source of data and tools to owners and managers of constructed facilities, community planners, policy makers, and other decision makers concerned with the disaster resilience of their constructed facilities and communities. The intended audience is the National Institute of Standards and Technology, the National Oceanic and Atmospheric Administration, the Department of Homeland Security's Science and Technology Directorate and Federal Emergency Management Agency, the U.S. Geological Survey, the National Science Foundation, insurance industry organizations, and standards and codes development organizations interested in developing metrics and tools for defining and measuring the disaster resilience of constructed facilities and communities.

Disclaimer

Certain trade names and company products are mentioned in the text in order to adequately specify the technical procedures and equipment used. In no case does such identification imply recommendation or endorsement by the National Institute of Standards and Technology, nor does it imply that the products are necessarily the best available for the purpose.

Cover Photographs Credits

Clockwise from the bottom right are as follows:
Path of named North Atlantic Tropical Cyclone Tracks in 2005, National Oceanic and Atomspheric Administration, National Hurricane Center; Tampa and vicinity map, 2009 National Geographic Society, i-cubed; Seismic Hazard Map of the Continental United States, nationalatlas.gov; DigitalVision Construction in Action clips gallery image used in compliance with DigitalVision's royalty free digital stock photgraphy use policy; and Microsoft Clip Gallery Images used in compliance with Microsoft Corporations non-commercial use policy.

Acknowledgements

The author wishes to thank all those who contributed so many excellent ideas and suggestions for this report. They include Mr. Stephen A. Cauffman, Deputy Chief of the Materials and Construction Research Division of the Building and fire Research Laboratory (BFRL) at the National Institute of Standards and Technology (NIST), and Dr. Jack Hayes, Director of the National Earthquake Hazards Reduction Program at BFRL/NIST, for their leadership, technical guidance, suggestions, and programmatic support. Special appreciation is extended to Dr. David T. Butry, Dr. Robert E. Chapman, and Dr. Harold E. Marshall of BFRL's Office of Applied Economics for their thorough reviews and many insights, to Mr. Douglas Thomas for his assistance in generating the charts used on the cover page, and to Ms. Carmen Pardo for her assistance in preparing the manuscript for review and publication. The report has also benefitted from the review and technical comments provided by Dr. Nicos Matrys of BFRL's Materials and Construction Research Division.

Table of Contents

Abstract iii
Preface v
Acknowledgements vii
1   **Introduction** 1
   1.1 Background 1
   1.2 Purpose 1
   1.3 Scope and Approach 2
2   **Disaster Resilience: A Survey of the Literature** 5
   2.1 Preliminaries 8
       2.1.1 Definitions 8
       2.1.2 What we are protecting 11
       2.1.3 Dimensions of Harm 13
       2.1.4 Disaster Cycle 14
   2.2 Response 15
       2.2.1 Observations from Research 16
       2.2.2 Ways to Improve Response 18
   2.3 Recovery 19
       2.3.1 Full Recovery 20
       2.3.2 "Instant Urban Renewal" 21
       2.3.3 Population Recovery 22
       2.3.4 Physical Infrastructure 22
       2.3.5 Economy 23
       2.3.6 Social Networks, Government Services and the Environment 25
       2.3.7 Simulating Recovery 25
       2.3.8 Ways to Improve Recovery 29
   2.4 Preparedness 30
       2.4.1 Warning Systems 31
       2.4.2 Evacuation 31
   2.5 Mitigation 33
       2.5.1 Individual Behavior 33
       2.5.2 Vulnerability 34
       2.5.3 Hazard Models and Damage Prediction 35
       2.5.4 Optimal Mitigation 36
       2.5.5 Optimal Insurance 38
   2.6 Data and Measurement 39
       2.6.1 Recovery Measurement 42
       2.6.2 Individual and Family 43
       2.6.3 Physical Infrastructure 43
       2.6.4 Economy 43
       2.6.5 Resilience 46

3   Summary and Recommendations for Further Research 49
    3.1 Summary 49
    3.2 Recommendations for Further Research 50
        3.2.1 Analysis of Key Structure Types 50
        3.2.2 Cost Analysis 51
        3.2.3 Analysis of Expected Property Losses 51
        3.2.4 Analysis of Expected Business Interruption Costs 52

Appendix A   Disaster Resilience: An Annotated Bibliography 53
Appendix B   Selected Databases, Software Tools, and Web Portals 93
Appendix C   Mathematical Appendix 99
    C.1 What is the Optimal Recovery Path from Disaster 99
    C.2 A Rationalization of the Multiplicative Resilience Index 101

References 105

List of Figures

Figure 1: Real U.S. disaster losses, in billions of dollars 5
Figure 2: U.S. deaths due to natural disasters 6
Figure 3: Hypothetical Trajectories of recovery 14
Figure 4: Disaster Cycle 15
Figure 5: Percent recovery versus time (in years) for different levels of destruction in a disaster 28
Figure 6: Time to recovery relative to initial capital and consumption 29
Figure 7: Conceptual degree of evacuation versus risk 32
Figure 8: Elements of mitigation, with dependencies 33
Figure 9: Time line of production for an individual firm following a disaster 44

List of Tables

Table 1: 2009 Swiss Re threshold for disasters 9
Table 2: MCEER PEOPLES framework 12
Table 3: What we want to protect 13

1 Introduction

1.1 Background

Despite significant progress in disaster-related science and technology, natural and man-made disasters in the United States are responsible for an estimated $57 billion (and growing) in average annual costs in terms of injuries and lives lost, disruption of commerce and financial networks, property damaged or destroyed, the cost of mobilizing emergency response personnel and equipment, and recovery of essential services. Natural hazards—including earthquakes, community-scale fires, hurricane-strength winds and hurricane-borne storm surge, and tsunamis—are a continuing and significant threat to U.S. communities. Preventing natural and man-made hazards from becoming disasters depends upon the disaster resilience of our structures and communities. While the overall goal is to improve resilience at the community scale, buildings play a critical role in assuring community resilience. Thus, measuring the resilience of buildings provides a foundation for the more complex problem of measuring the resilience of communities.

Measuring the resilience of structures and communities to natural and man-made hazards poses difficult technical challenges due to a lack of adequate: (1) understanding of the natural processes that create hazards to the built environment and information relative to such hazards for use by design professionals, standards developers, and emergency managers; (2) predictive technologies and mitigation strategies to improve the performance of complex structural systems and means to transfer the results of research into practice and to promote risk-informed behavior; and (3) standard methods to assess the disaster resilience of structures and communities for use in making disaster preparedness and mitigation decisions. Buildings are systems of systems with many interactions and interdependencies among these systems. The performance of the structural system can have a direct impact on the performance of other systems (e.g., architectural, mechanical, electrical, etc.), in addition to utility infrastructure that can be housed inside, atop, or below the building. Thus the resilience of the structure has a substantial influence on the resilience of the building as a whole and in turn on the community. Although the focus of this report is on a critical review of and commentary on resilience-related literature, it serves to strengthen the linkage between establishing technically valid measures of resilience at the level of the individual building and at the level of the community.

1.2 Purpose

Although there is a great deal of high-quality information available on resilience-related topics—hazard assessment, vulnerability assessment, risk assessment, risk management, and loss estimation—as well as disaster resilience itself, there is no central source of data and tools to which the owners and managers of constructed facilities, community planners, policy makers, and other decision makers can turn for help in defining and measuring the resilience of their structures and communities. The purpose of this document is to provide a survey of the literature and an annotated bibliography of printed and electronic resources that serves as that central source of data and tools to help readers develop methodologies for defining and measuring the disaster resilience of their structures and communities.

The report covers resilience-related topics at two different levels: (1) individual constructed facilities and correlated collections of constructed facilities that form a network (e.g., hospitals) and (2) community/regional scale frameworks (e.g., business and economic relationships, population and employment demographics). Thus, the first level focuses on physical infrastructure, whereas the second takes a broader look at how the physical infrastructure interacts with other activities that collectively define modern communities. The reason for taking this approach is to establish a foundation for developing methodologies for defining and measuring the disaster resilience of structures. This step is especially important because physical infrastructure enables the community to function as a place of employment, a window to the regional and national economy, and a home for individuals. Developing better metrics and tools for defining and measuring the resilience of structures is an important step in meeting the challenge of measuring disaster resilience at the community scale.

The report is being published both as a printed document and as a Web-based publication. Both documents are intended as a ready reference for researchers, public and private sector decision-makers, and practitioners—owners and managers of constructed facilities—concerned with defining and measuring disaster resilience. In the Web-based version downloadable from the NIST Website, all Web links are active as of the print date, enabling readers to browse documents and data sources. Furthermore, many of the Web links permit documents and data files to be downloaded for future reference and use.

1.3   Scope and Approach

The report contains two chapters and two appendices in addition to the Introduction. Chapter 2 and Appendix A are the core components of the report.

Chapter 2 is a survey of the literature on disaster resilience. The chapter begins with a discussion of the key concepts—hazards, vulnerability, losses, and disasters—that feed into any methodology for defining and measuring disaster resilience. The discussion then focuses on the various definitions of disaster resilience that are found in the literature. For the purpose of setting the scope of the literature review found in this report, disaster resilience is defined as the ability to minimize the costs of a disaster, return to the *status quo*, and to do so in the shortest feasible time. A conceptual model of the "disaster cycle" is then introduced. The disaster cycle model consists of four stages: (1) response; (2) recovery; (3) preparedness; and (4) mitigation. The first two stages are "reacting to" a disaster, whereas the second two are "preparing for" a disaster. The conceptual model provides an ideal framework for highlighting the interactions between the four stages and how decision-making at the individual asset and community levels differ. The chapter then describes data and measurement issues associated with loss estimation. These include critical analyses of case studies, insurance claims, direct measurement of losses, and survey-based methods. The discussion covers both direct losses due to damages to buildings and other constructed facilities and indirect losses due to business interruption and other second-order, disaster-related effects. The chapter concludes with a mathematical treatment of two resilience-related topics—the optimal recovery path following a disaster and the theoretical basis for a multiplicative resilience index.

Chapter 3 provides a summary and recommendations for further research. Four recommendations for further research are put forward with an aim of strengthening the linkage between defining and measuring disaster resilience at the level of individual structures and at the community level.

Appendix A presents an annotated bibliography on disaster resilience. The appendix begins with a synopsis of the key reference documents that provide the basis for much of the literature on resilience-related topics. References are listed in alphabetical order by author, where the author may be a person, a company, an organization, or a government entity. An abstract is provided for each reference that summarizes the salient points of the reference.

Appendix B presents an annotated bibliography of selected databases, software tools, and Web portals. An abstract is provided for each reference that summarizes the salient points of the reference. A URL is provided whenever a reference is available in electronic format.

2 Disaster Resilience: A Survey of the Literature

Disaster losses have been increasing exponentially, and many researchers argue that the increasing trend is unsustainable (e.g., see Mileti 1999; White, Kates, and Burton 2001), and that eventually the costs of disasters will become more than we can afford. Real U.S. Disaster losses since 1960 are shown in Figure 1. The trend, shown in the graph, shows that disaster losses are increasing at a 4.8 % annual rate. A similar trend is visible in international data as well. One factor influencing the results in the figure is that the data are biased down.[1] The bias tends to be stronger the older the data.

Figure 1: Real U.S. disaster losses, in billions of dollars

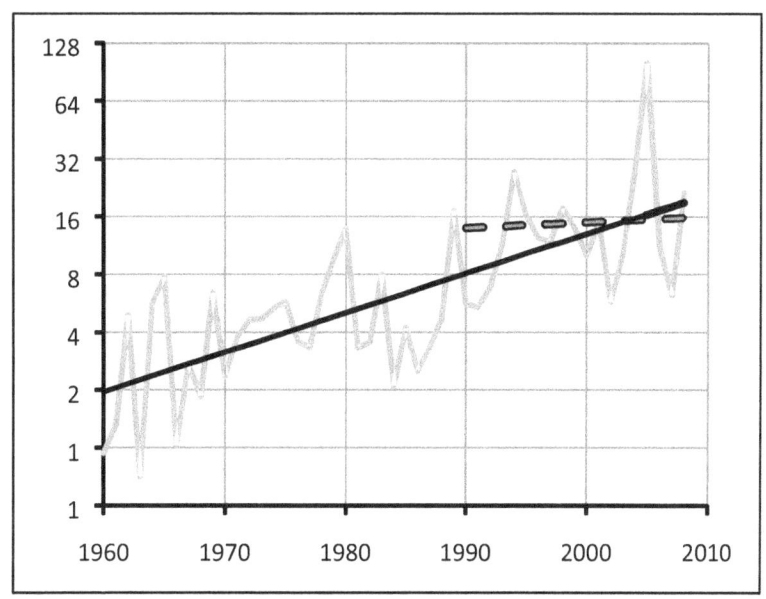

Source: Data drawn from SHELDUS.

A number of researchers have investigated the contributors to this trend (Collins and Lowe 2001; Cutter and Emrich 2005; Dlugolecki 2008; Landsea et al. 2003; Miller, Muir-Wood, and Boissonnade 2008; Pielke and Downton 2000; Pielke and Landsea 1998; Vranes and Pielke 2009). These authors have attempted to determine how much of the trend is due to increases in population and wealth, spatial distribution of assets, and climate change. Confounding the analysis is the fact that there was a lull in Atlantic hurricane activity between about 1970 and 1990 (Landsea et al. 2003). Most analyses have found that damage trends are flat or down after adjusting for population and wealth. Pielke and Downton (2000) confirmed those results but also found that damaging floods are most strongly correlated with the number of 2-day heavy rainfall events, and that such events are increasing.

---

[1] That is, the values in the figure are probably lower than the true damages.

Further analysis reveals that the trend since the early 1990's is essentially flat (see Figure 1). Similar results are present in international data (not shown). Disaster loss data from the Center of Research on the Epidemiology of Disasters and from Swiss Re both have a break in trend and level of damages somewhere around 1989, with the trend flattening. In the Swiss Re data the structural break occurs in insured losses from both weather-related disasters and in man-made disasters. It is not clear at this point whether the flattening of the trend is due to changes in frequency and severity of hazards, improved mitigation, changes in data collection and reporting, or statistical chance.

In contrast to property losses, deaths and injuries due to disasters have been flat for decades (see Figure 2). Similar results hold for international losses (not shown). Analysis of data from Swiss Re finds no significant trend in deaths internationally due to natural hazards. However, there is a significant upward trend internationally in deaths due to man-caused disasters. No similar data set for man-caused disasters was readily available for the U.S..

Figure 2: U.S. deaths due to natural disasters

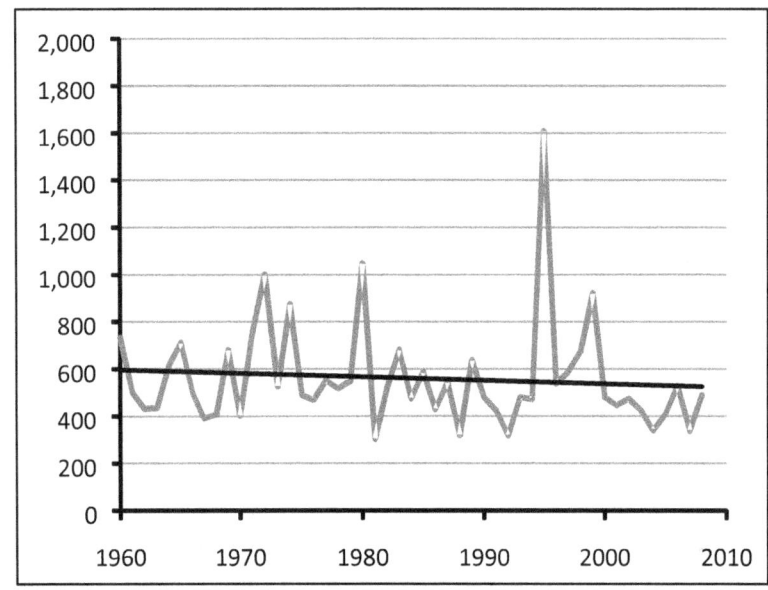

Source: Data drawn from SHELDUS.

Different hazards cause property damage versus deaths. The property loss record in the U.S. is dominated by hurricanes, flooding and other coastal hazards (for example Stuart Miller et al. 2008). Deaths are dominated by heat, thunderstorms and lightning, and severe winter weather (Borden and Cutter 2008).

Data quality issues will be discussed in more detail later, but two items from the above figures illustrate some of the problems with disaster-loss data. Total damages for 1992 in Figure 1 are about $7 billion. However, in 1992 Hurricane Andrew struck south Florida and did in excess of $30 billion in damages. Similarly in Figure 2, deaths from Hurricane Katrina do not appear.

Neither omission changes significantly the results discussed above, but they do illustrate that data quality is an issue.

Disaster losses are highly skewed. For example, the top 10 % of damaging floods do approximately 80 % of the property damage.[2] It is easy to pick the Northridge Earthquake and Hurricane Katrina out of the graph in Figure 1. The spikes in the number of deaths in Figure 2 are almost all individual heat waves, including 1980, 1995 and 1999. After normalizing for population growth, wealth and inflation, the disaster record suggests that hurricanes are the most destructive type of U.S. disaster, with about $1.05 trillion in aggregate normalized losses from hurricanes since 1900 followed by earthquakes with about $432 billion in aggregate normalized losses since 1900 (Vranes and Pielke 2009). However, the largest events for hurricanes and earthquakes are of comparable size, with the largest U.S. disaster in terms of normalized damages probably being the 1906 San Francisco Earthquake.

A number of groups have been warning that the U.S. is becoming more vulnerable to disasters (e.g., Burby 1998, Berke et al. 2008). First, people have been moving into areas with greater exposure to natural disasters. The percentage of the population living in areas with high exposure to hurricanes has been steadily increasing.

Second, many warn that U.S. infrastructure has become inherently more vulnerable. For example, the American Society of Civil Engineers in its 2009 Report Card for America's Infrastructure (2009) gives most infrastructure systems a 'D' grade, and warns of dire consequences if infrastructure is not improved. For example, with regard to waste water, it warns that "the nation's wastewater systems are not resilient in terms of current ability to properly fund and maintain, prevent failure, or reconstitute services." Furthermore "if the nation fails to meet the investment needs of the next 20 years, it risks reversing public health, environmental, and economic gains of the past three decades." [3]

Because of this, the National Science Technology Council developed a set of Grand Challenges for Disaster Reduction (2005) whose objective was to increase disaster resilience and reduce harm from natural and man-made disasters. The grand challenges were:

Grand Challenge #1—Provide hazard and disaster information where and when it is needed.

Grand Challenge #2—Understand the natural processes that produce hazards.

Grand Challenge #3—Develop hazard mitigation strategies and technologies.

---

[2] The results from Vranes and Pielke (2009) suggest that damages from earthquakes are even more highly skewed than from hurricanes or floods.

[3] There are two main problems with the report card. First there are no clear criteria for the grades they give; so there is no way to independently evaluate the report and no way to determine what is required to raise the grades. Second their grades do not appear to be performance based.

Grand Challenge #4—Recognize and reduce vulnerability of interdependent critical infrastructure.

Grand Challenge #5—Assess disaster resilience using standard methods.

Grand Challenge #6—Promote risk-wise behavior.

This report is a step in the process of meeting several of these grand challenges, but primarily in meeting Grand Challenge 5.

2.1 Preliminaries

2.1.1 Definitions

A *Disaster* according to the dictionary[4] is "a sudden or great misfortune." Researchers have defined disaster in a number of different ways. Rubin (1985) defined disaster as "an unscheduled, overwhelming event that causes death, injury, and extensive property damage." Mileti (1999) quotes several definitions of disaster, including "a breaking of the routines of social life in such a way that extraordinary measures are needed for survival," and "disruptions in cultural expectations that cause a loss of faith in the institutions that are supposed to keep hazards under control."

Fritz (1961) developed one of the earliest and most influential definitions, when he defined a disaster as:

> an event, concentrated in time and space, in which a society, or a relatively self-sufficient subdivision of a society, undergoes severe danger and incurs such losses to its members and physical appurtenances that the social structure is disrupted and the fulfillment of all or some of the essential functions of the society is prevented

Swiss Re (2010) operationally defines a disaster as any event in which insured losses exceed the thresholds listed in Table 1. Swiss Re adjusts the thresholds annually. Operational definitions of this sort are common, but are not consistent across databases. For example, the SHELDUS database now includes all geologic and weather-related hazards since 1970 that have killed at least one person, and Property Claims Services defines a disaster as an "event that causes $25 million or more in direct insured losses to property...."

---

[4] **Webster's Third New International Dictionary, Unabridged**. Merriam-Webster, 2002. http://unabridged.merriam-webster.com (18 Jun. 2010).

Table 1: 2009 Swiss Re threshold for disasters

Type of Disaster		Losses ($M)
Insured Claims	Maritime Disasters	17.1
	Aviation	34.3
	Other Losses	42.6
or Total Economic Losses		85.2
or Casualties	Dead or Missing	20
	Injured	50
	Homeless	2,000

For the purposes of this document, a *Disaster* is defined as any event appearing in a disaster database, with the understanding that the boundaries of the definition are very fuzzy at the lower end.

A *Hazard* is a natural or man-made phenomenon capable of inflicting harm on communities. A disaster is the intersection of some hazard with a vulnerable community (National Science Technology Council 2005). In the discussion *Hazard* is used to mean any phenomenon capable of causing harm, whether it does so or not, and *Disaster* to refer to any situation where harm has been caused on a wide scale in a community.

Resilience has a wide variety of definitions as well. The basic idea is that "a community's hazard resilience is ... its ability to absorb disaster impacts and rapidly return to normal socioeconomic activity" (Lindell in NEHRP 2010). In general, definitions for resilience fall into two broad categories. The first are outcome-oriented definitions, and the second are process-oriented definitions.

Outcome-oriented definitions define resilience in terms of end results. An outcome-oriented definition would define resilience in terms of degree of recovery, time to recovery, or extent of damage avoided. For example, the Heinz Center (2000) defines a "disaster resilient community" as "a community built to reduce losses to humans, the environment, and property as well as the social and economic disruptions caused by natural disasters." Similarly Bruneau et al (2003) define resilience as "the ability of social units...to mitigate hazards, contain the effects of disasters when they occur, and carry out recovery activities in ways that minimize social disruptions and mitigate the effects of future [disasters]."

Process-oriented definitions have become preferred by disaster researchers from the social sciences. For example, Comfort (1999) defines resilience as "the capacity to adapt existing resources and skills to new situations and operating conditions." Norris et al. (2008) define resilience as "a process linking a set of adaptive capacities to a positive trajectory of functioning and adaptation after a disturbance."

Manyena (2006), arguing forcefully for a process-oriented definition of resilience, says that

> [t]he danger of viewing disaster resilience as an outcome is the tendency to reinforce the traditional practice of disaster management, which takes a reactive stance. Disaster management interventions have a propensity to follow a paternalistic mode that can lead to the skewing of activities towards supply rather than demand. Activities such as community capacity building, mitigation and emergency preparedness planning, which impact greatly on response and recovery operations, may be neglected. … Outcome-oriented disaster resilience programmes are inclined to adopt command and control styles that risk preserving the status quo, and which might entrench exclusion, and take attention away from the inequality, oppression and entitlement loss that results in cases of proneness to insecurity and disaster.

Furthermore he argues that

> Viewing disaster resilience as a deliberate process (leading to desired outcomes) that comprises a series of events, actions or changes to augment the capacity of the affected community when confronted with … shocks and stresses, places emphasis on the human role in disasters. Disaster resilience is seen as a quality, characteristic or result that is generated or developed by the processes that foster or promote it. Put differently, resilience is not a science nor does it deal with regularities in our experience, but rather, it is an art that addresses singularities as we experience them. For instance, recognising the human role in disasters, taking responsibility for action, having a disaster plan, building capabilities to implement the plan, purchasing insurance and sharing information on recovery priorities are steps that can enhance the resilience and hence the ability … to deal with [disasters].

Basically, Manyena is arguing three things. First, he argues that by focusing on desired outcomes, communities tend to overlook important measures that would improve their ability to withstand and respond to disasters. Second, there are some process-oriented items that are valuable in their own right. Here would belong "recognising the human role in disasters, [and] taking responsibility for action." Third, is a debate about the end point. Assume serious damage has occurred. Resilience would entail a prompt return to some improved state, but what state? A return to the *status quo ante* is never completely possible. However, a return to a state as good as the *status quo ante*, with a minimum of change presumably is. Manyena essentially argues that the *status quo ante* was flawed, and the disaster should be seen as an opportunity to fix some of the flaws. For example, he argues that the end-point for resilience should include correcting problems with "inequality, oppression and entitlement loss."

An outcome-oriented definition better suits our purposes than a process-oriented one. Process in dealing with disasters is only valuable to the extent that it serves specific desirable outcomes: saving lives, preventing the destruction of property and the environment, maintaining the flow of valuable goods and services, etc. A process-oriented definition begs the key question of what process? That is a question that cannot be answered without reference to the desired outcomes. An outcome-based definition and measure would enable us to more rigorously test processes to see which ones are effective at minimizing the destruction associated with disasters.

Therefore for this report *Resilience* is defined as the ability to minimize the costs of a disaster, to return to a state as good as or better than the *status quo ante*, and to do so in the shortest feasible time. As such, greater resilience entails a return to some 'normal' as soon as possible, and the best 'normal' reasonably achievable. Note that this is general enough that it leaves room for a substantial amount of community reorganization or redistribution if the community desires it. However, at the same time it does not take it into account. Reorganization of the community may very well be desirable and desired by the citizens of the community, but community redesign and improvement are different from resilience.

Norris et al. (2008) make a distinction between resilience and *Resistance*. In their terminology resilient communities and people bounce back from disasters,[5] while *Resistant* communities and people do not suffer harm from hazards in the first place. In this paper *Resistance* is used to mean the ability to withstand a hazard without suffering much harm. *Resilience* in this paper will include resistance but will also include the ability to recover after suffering harm from a hazard.

2.1.2 What we are protecting

In order to effectively measure and improve resilience, we need to define what bad thing it is we are trying to avoid. That is, we need to know what it is we are trying to protect. In the literature that has usually been left implicit. Generally the focus is on protecting either lives or property.

The Heinz Center (2000) divided losses into four categories: the built environment, the business environment, the social environment, and the natural environment. Losses to the built environment include destruction of houses and businesses as well as destruction of lifeline infrastructure. Losses to the business environment included what are usually called "indirect" business interruption losses as well as losses of things like inventory, tools, and equipment.

---

[5] ...to some extent. Their definition is process oriented and so is deliberately vague about how much the resilient community or person bounces back.

Losses to the social environment included deaths, injuries, hedonic losses due to life and community disruption, and lost wages. Losses to the natural environment included among other things crop losses.

The MCEER (2010) recently developed a draft framework for measuring disaster resilience. There are seven elements in the framework (represented by the acronym 'PEOPLES') which represent both what we want to protect and resources that provide protection. The framework to some extent seems to confound the roles of 'protected' and 'protection' in the sense that each element in the framework fills both roles. The framework does not make the respective roles of 'protected' versus 'protection' clear. Here we are interested in the elements of the framework as things we want to protect.

The elements in the framework are listed in Table 2. Most elements are fairly self-explanatory. However, Lifestyle and Community Competence and Social-Cultural Capital are not clearly separated.

Table 2: MCEER PEOPLES framework

Population and Demographics
Environmental/Ecosystem
Organized Governmental Services
Physical Infrastructure
Lifestyle and Community Competence
Economic Development
Social-Cultural Capital

For the purposes of this paper a slight modification of the PEOPLES framework is used. The list used in this paper is shown in Table 3. Protecting people means, among other things, preventing deaths and injuries and preventing people from being made homeless. Protecting physical infrastructure means limiting damage to buildings and structures, including most lifeline infrastructure. Protecting the economy means preventing job losses and business failures, and preventing business interruption losses. Protecting key government services includes (among other things) ensuring that emergency services are still functioning. Protecting social networks and systems includes (among many other things) ensuring that people are not separated from friends and family.

Table 3: What we want to protect

People
Physical Infrastructure
Economy
Key Governmental Services
Social Networks and Systems
Environment

2.1.3 Dimensions of Harm

In measuring the harm done by a disaster there are a number of dimensions that need to be considered.

First, we need to consider both *direct* and *indirect* damages. *Direct* damages are those that are directly caused by the hazard. For example, this would include people killed or injured in an earthquake, buildings destroyed in a hurricane, forest area burned in a wildfire, etc. Direct damages are what the traditional measures of harm tend to measure.

However, there are also *indirect* damages. Indirect damages are those damages that are caused, not by the hazard itself, but as a 'ripple effect' from the initial damages. For example, this would include business interruption costs due to power outages, deaths resulting from a lack of medical services, etc. Indirect effects need not always be negative. For example, Hurricane Hugo devastated the Frances Marion National Forest, thus doing long-term harm to the forestry industry in South Carolina. However, the destruction in the Forest produced a short-run boom as salvage operations cleaned up after the hurricane (Guimaraes, Hefner, and Woodward 1993).

Trajectory of recovery also matters to the total amount of damages (see Figure 3). If direct harm is the same for two disasters, but recovery time is longer for one (for example, the power is out longer, or people are unable to return to their homes for a longer time) then the longer recovery time results in greater total harm. Similarly, a disaster where recovery is complete will suffer less harm than one where recovery is only partial. In some cases recovery may be (temporarily) more than complete (see Figure 3).

Figure 3: Hypothetical Trajectories of recovery

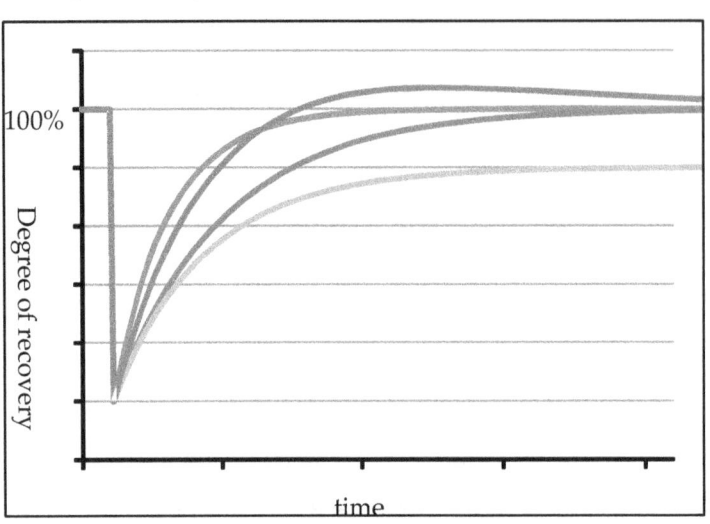

As a general rule we tend to measure harm at the local and regional level. However, effects may spill over into the economy as a whole. First, few if any regions are truly autonomous. An earthquake that interrupts business in Los Angeles will have secondary effects throughout the country. Businesses in Los Angeles will have customers in Denver and suppliers in Seattle. Furthermore, when recovery is incomplete, one region's losses may be another region's gains. Population losses to New Orleans due to Hurricane Katrina were population gains for Houston and Memphis.

2.1.4 Disaster Cycle

Most recent researchers use the 'disaster cycle' as a filter for understanding and evaluating response to hazards. By portraying it as a cycle, we convey the idea that in recovering from one disaster we are preparing for the next one.

Different researchers use different versions of the disaster cycle depending on their purposes and objectives. For example, MCEER (2010) uses a disaster cycle with six stages. Rubin (1991) on the other hand divides the disaster cycle into four stages. This report uses the disaster cycle from Rubin (1991), shown in Figure 4.

The stages in the disaster cycle can be thought of as falling into two groups. Response and Recovery form the first group, while Mitigation and Preparedness form the second group. Response and Recovery occur in the immediate aftermath of a disaster and are generally aimed at dealing with the disaster that has just hit. Mitigation and Preparedness occur mainly during normal circumstances and are largely aimed at preparing for the next disaster.

Figure 4: Disaster Cycle

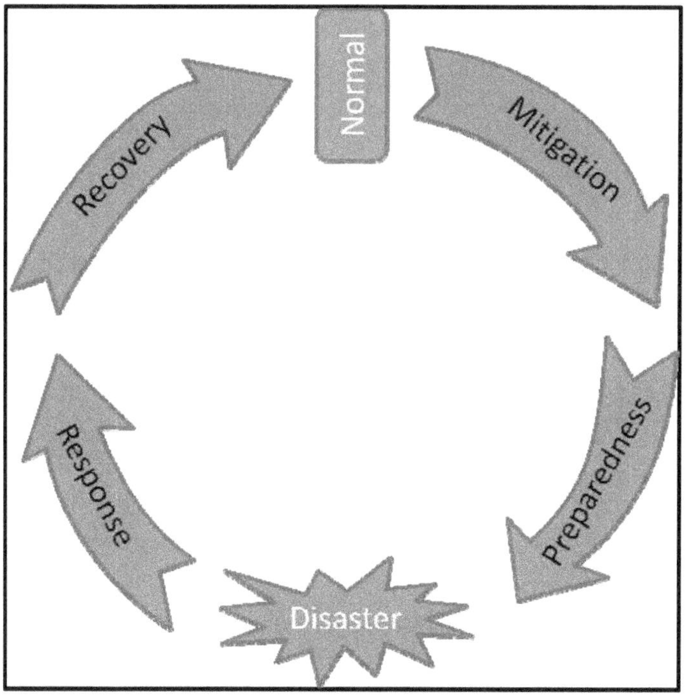

2.2 Response

Response includes the period during and immediately following a disaster in which efforts are directed at limiting the harm done by the disaster (Haas, Kates, and Bowden 1977). The NFPA (2007) defines *response* as the "immediate and ongoing activities, tasks, programs, and systems to manage the effects of an incident that threatens life, property, operations, or the environment." This includes measures like evacuation, mass emergency sheltering and feeding, emergency response, search and rescue, and initial clearing of the main arteries. During the response period normal social and economic activities are stopped.

Harrald, (2006) divides the response period into four phases.

- An initial response phase. This is conducted by local resources while external resources are being mobilized.

- An integration phase. Once external resources have arrived, time is required to structure these resources into a functioning organization.

- A production phase. During the production phase, the response organization is fully productive, delivering needed services as a matter of routine.

- A demobilization phase.

2.2.1 Observations from Research

There have been many case studies on response, among which are Comfort (1999), Comfort and Haase (2006), Quarantelli (1982, 1983), and Wenger et al. (1986, 1989) (See Quarantelli 1989 for a more complete bibliography). There are a number of very consistent results regarding the response period from the case studies. Unfortunately, it is also very hard to determine from the literature what measures can be taken by communities to improve response.

Virtually all studies agree on a number of things (see for example Perry and Lindell 2007). Citizens in disasters do not panic; they do not react with shock or passivity. People are almost always more prosocial after disasters, and act rationally given their understanding of the situation. Similarly, "people with officially-defined disaster roles will execute those roles…" but will worry about their families until they find out how they are doing. That is, they do not abandon their posts (Perry and Lindell 2007).

"Uninjured victims are usually the first to search for survivors, care for those who are injured, and assist others in protecting property." In fact many report that the majority of search and rescue is performed by bystanders rather than professional emergency responders (ibid.).

Very few people use mass emergency shelters. In general 5 % - 15 % of evacuees stay in emergency shelters (Tierney, Lindell, and Perry 2001). Most people who evacuate stay in hotels or with family or friends. Emergency shelters also tend to empty out very quickly (Quarantelli 1982).

In relatively small disasters, there are far more resources than can be used. In the initial stages of a disaster the scale of the disaster it often not clear, so there is a tendency to dispatch more resources than turn out to be needed. In addition, many emergency response units self-dispatch. For example, in the well-studied 1974 Xenia Ohio tornado, fire units from Dayton traveled the 15 miles to Xenia to assist in search and rescue without any requests for assistance from Xenia (Quarantelli 1982).

Even in larger disasters resources are not usually scarce. For example, in Hurricane Hugo in response to the persistent national appeals by the mayor of Charleston, South Carolina for food and clothing, the region was inundated by donations, many of which simply had to be thrown away (Rubin 1991). The biggest resource problems in both small and large disasters tend to be getting the *right* resources[6] and distributing the available resources to those who need them.

Local emergency-response agencies will be the first official responders to disasters. In small disasters they may be the only such responders. In large disasters they typically provide the sole

---

[6] Rubin (1991) reports in the Charleston, SC case that the food and clothing donations "contributed to significant roadway congestion and … a flood of telephone calls that diverted disaster workers' attention." In addition, it "got in the way of shipments of chainsaws and other needed goods."

emergency response for hours to days after the disaster. Generally they respond in ways very similar to their behavior in normal situations (Quarantelli 1983; Wenger et al. 1989).

Police and Fire departments do not reach much beyond their regular duties and day-to-day experience in disasters. Basically, they do what they always do. Firefighters put out fires, engage in search and rescue, and provide emergency medical treatment (at least, those departments that provide EMT services normally) and then go back to the station. Police departments engage in search and rescue and provide site control. Emergency medical personnel stabilize victims and transport them to the nearest hospital. All these agencies and personnel mostly do their job, with very little modification, the way they always do it.

This has prompted a fair amount of criticism for the agencies from researchers for failing to adapt to the environment of the disaster. Disasters are different. As already discussed, many EMS tasks are carried out by people outside the traditional emergency system. In larger disasters agencies may have to temporarily ration services. Interagency coordination becomes much more important. And often changes in routines can help alleviate some of the problems that develop.

On the other hand, there are benefits to sticking mostly to what they do normally. In many of the case studies (which were all small disasters) the coordination problems that could have developed generally did not precisely because the agencies basically limited themselves to their normal roles. Furthermore, since agencies rarely went beyond their training, the work was carried out expeditiously and well.

One partial exception to this was emergency medical care. The main study on this is Quarantelli (1983), which is almost 30 years old. The much more recent paper by Auf der Heide (2006) repeats the same lessons.

The main objective of all participants in emergency medical treatment is to get victims to the nearest hospital as quickly as possible. Triage in the field simply was not performed. Some 40 % of patients were not transported to the hospital by ambulance (Auf der Heide 2006 suggests that the number tops 50 %). As a result less-injured patients tend to arrive first at the hospital.

Communication was poor to nonexistent. What information hospitals had generally came from ambulance drivers and early-arriving patients. So, hospitals did not generally know what to expect. As usual throughout the emergency response system they usually prepared for a much bigger disaster than they actually ended up facing.

Due to the tendency to transport patients to the nearest hospital, one hospital was frequently overwhelmed with patients, while other nearly hospitals had few if any patients. Occasionally this ended up costing lives because hospitals drew personnel from other critical areas to deal with the large patient influx.

An important observation is that in disasters hospitals just do not keep good records of patients, injuries, and treatments, especially for those with minor injuries. So, data on injuries is likely to be unreliable.

The big problems in response, through many disasters, over many years, in many countries, were communication, task assignment and coordination, and authority relationships. Essentially, you had independent units working, not talking to each other, and no one (that everyone agreed upon) in charge (see for example Comfort 1999; Quarantelli 1983; Wenger et al. 1986).

In small disasters this was not generally a big issue since interagency coordination was either not needed or could be handled implicitly by assuming the other agencies would fulfill their traditional roles. However in larger disasters this becomes a major problem. When outside agencies are dispatched (or self-dispatch) coordination with the local agencies is often poor. So, for example, fire fighting units may sit unused because no one says 'you are needed over there.'

Comfort (1999) studied communications extensively in her work on disaster response internationally. In all her case studies communications were a serious problem. In general she found that the better communications were handled the better the response was.

2.2.2 Ways to Improve Response

The research has been much less effective at finding ways to improve response. Some of the regularities that researchers have identified in response suggest ways (in theory at least) that response can be improved. The big observation of all the researchers cited is that good interagency communications is important to good disaster response. The importance of good communications increases as the size of the disaster response increases—and hence the number of different agencies increases. Unfortunately that seems to be difficult to do. As Wenger et al. (1986) said, "[r]esearchers have noted these issues for the past two decades. Even in the face of concerted efforts to improve planning, they continue to plague disaster response."

Disaster planning correlates poorly with good disaster response (Tierney et al. 2001; Wenger et al. 1986). Wenger, et al. (1986), in their case studies, found some evidence that planning helped, but even in their work the correlation was poor. The best predictors of effective response were not having a plan, a local emergency management office, or anything else. The best predictor was experience. The communication problems mentioned still occur, but they were less severe and less common.

Harrald (2006) argued that both agility and discipline are critical factors for success in disaster response. Agility is needed because much of what happens in a disaster will be unanticipated. Discipline is needed because it increases the effectiveness of the actions taken. He also argues that they are not incompatible.

A number of authors have criticized the various incident management systems that have been used in this country, including the National Incident Management System (NIMS). Harrald (2006) argued that the NIMS is rigid and tends to sacrifice agility. Wenger, et al. (1989), looking at an earlier version of NIMS, argued that the Incident Management System was in effect a Fire Management System. In the case studies they evaluated,[7] the Incident Management System was used exclusively by fire departments. When multiple agencies were involved, police departments and other groups were not a part of the Incident Management System.

2.3 Recovery

> ...ὁμοίως σμικρὰ καὶ μεγάλα ἄστεα ἀνθρώπων [ἐπέξειμι]. τὰ γὰρ τὸ πάλαι μεγάλα ἦν, τὰ πολλὰ σμικρὰ αὐτῶν γέγονε· τὰ δὲ ἐπ᾽ ἐμεῦ ἦν μεγάλα, πρότερον ἦν σμικρά. τὴν ἀνθρωπηίην ὦν ἐπιστάμενος εὐδαιμονίην οὐδαμὰ ἐν τὠυτῷ μένουσαν, ἐπιμνήσομαι ἀμφοτέρων ὁμοίως. Herodotus, Histories, 1.5.3-4.

> I will treat cities with both small and large populations alike. For many of the cities which were large in ancient times now have become small. Those that are large in my time formerly were small. Knowing that good fortune in a place does not last, I will treat them both alike.

Recovery is the time period in which activities and programs are pursued with the objective of returning conditions to an acceptable level (National Fire Protection Association 2007). During the recovery period, rubble is cleared, people are provided with interim housing, normal social and economic activities are resumed, and the area is rebuilt (Haas et al. 1977).

Kates and Pijawka (1977), part of the larger work of Haas et al. (1977), break recovery down into four different sub-periods: an Emergency Period, a Restoration Period, a Replacement Reconstruction Period, and a Commemorative, Betterment, Developmental Reconstruction Period. Their Emergency Period is approximately comparable to the Response period in section 2.2. During the Restoration period infrastructure is repaired and normal activities are resumed. The Replacement Reconstruction Period rebuilds capital stock to predisaster levels. Finally the Commemorative-Betterment period is when memorials are built and attempts are made to improve the community. They argued that the length of time for each period was fairly predictable, with each period lasting about 10 times as long as the previous period (except for the Commemorative-Betterment Period).

The periods were overlapping, with restoration activities often begun before the emergency activities were complete, and replacement reconstruction begun before restoration was fully complete. As a general rule recovery was faster for the wealthy than the poor. So Replacement

---

[7] Their case studies date from the 1980's. It is not clear from the literature how much of their observations still hold.

may be well underway in one segment of the city even while Restoration is ongoing in a different portion of the city.

Many more recent researchers have criticized this framework (for example Berke, Kartez, and Wenger 2008; Miles and Chang 2003). However, much of the criticism is really a result of the more recent researchers being interested in different questions. The more recent researchers are often interested in differences in impact and recovery far more than they are interested in the fact of recovery. As Miles and Chang point out "this more recent literature has been concerned with disparities and equalities in recovery, and with conceptualizing disaster recovery as a social process involving decision-making, institutional capacity, and conflicts between interest groups."

The results of Kates and Pijawka were based on four case studies of major disasters throughout the world. The best-studied case was the 1906 San Francisco Earthquake. As such the regularity of the lengths of recovery periods is not well supported. The model itself is a useful way of thinking about some aspects of recovery, but, as the criticism suggests, its usefulness decreases as the focus shifts away from restoration of the physical city.

Berke et al. (2008) propose a model of 'recovery' intended to compete with Haas et al. However, it is really a model of aid. Their model is further hampered by their exclusive focus on government action — the private sector simply does not exist in their model.

2.3.1 Full Recovery

Usually we think of *full recovery* as a return to conditions as good as those that would have obtained had the disaster never occurred. Operationally, this is often defined as a return to conditions as good as or better than the *ex ante* state (e.g., Chang 2010).

By the latter definition, communities almost always fully recover. Kates and Pijawka (1977) analyze the time required for full recovery of population for cities hit by disaster between 1200 and 1800. These include a number of cities that lost more than 50 % of their population in disasters (often wars). Nearly all the cities recovered. Vigdor (2008) does a similar analysis for modern disasters and finds the same result. He cites Hamburg where nearly half the city's housing was destroyed in allied bombing, yet the city's population had exceeded its prewar level by the early 1950's. The only modern example he could find where a city did not recovery was Dresden after World War II. Similarly, Quarantelli (1989) notes that "disasters are not new phenomena. But more important, recovery from them [is] also not new. In fact, there is almost always recovery from disasters."

Wright et al. (1979) analyzed the long-range effects of disasters on communities. They analyze all disasters to hit the U.S. in the 1960's, by county and census tract. As usual the overwhelming majority of the disasters are small: the median tornado, for example, killed no one and caused less than $500 000 in damage. Based on their database, they could find no evidence of long-term impact from a disaster on either the county or census tract it was in.

Vigdor suggests (and most researchers agree) that cities recover fully unless they are already in decline. His example of Dresden supports this since Dresden was already experiencing a population decline before the war. Vigdor argues on that basis that New Orleans likely will not recovery fully since it was already in a long-term decline.

Kobe does not seem to have recovered fully since the Great Hanshin Earthquake in 1995, and there is no sign of earlier decline (Chang 2010). In the case of Kobe, its port was an important economic foundation, and the port was completely destroyed. By the time the port was rebuilt three years later, business had moved elsewhere. Container traffic had been a major part of the port's business and had been growing before the earthquake. The earthquake was associated with a severe decline in container traffic, and ten years later container transshipments are down 90 %.[8]

2.3.2 "Instant Urban Renewal"

We have so far assumed that the objective of recovery is return to *ex ante* conditions as far as possible. For the most part we will continue to make that assumption in the remainder of this report. However, it is not the only possible objective, and many researchers argue that disasters are opportunities to reengineer the social environment in ways that improve resilience and social amenity. As a dramatic example, Foster and Giegengack (2006) essentially argue that New Orleans is irremediably vulnerable to hurricanes and other coastal disasters and should largely be abandoned. Mileti (1999) spins an imaginary scenario in which in the aftermath of a major hurricane hitting south Florida, the people decide to dramatically reduce the population of south Florida, largely abandon the coastline, and prohibit private automobiles. Proponents of using disasters as an opportunity for "instant urban renewal" on a smaller scale include Berke et al. (2008), Burby (1998), Godschalk et al. (1999), and Rubin (1985), among others.

There can almost never be a complete return to conditions *ex ante*. The deceased will never come back. Capital destroyed may be replaced, but only by either reducing consumption or foregoing capital investment elsewhere. In major disasters the layout of the city is almost always altered permanently (as discussed in more detail below). Furthermore, a jurisdiction hit by a disaster may decide to use the opportunity for redesign of some aspects of the community. Any measure of recovery will need to be robust enough to allow for these possibilities.

---

[8] Similarly, looking at ancient disasters, a number of cities lost population and took many centuries to recover (Data on population taken from Chandler (1987)). Baghdad was sacked in 1258 and took many centuries to recover. Similarly, Rome and Constantinople suffered major declines after being sacked. Rome did not return to its peak population for more than 1,000 years.

However, in each of these cases, the 'disaster' was as much political and economic as military. Rome and Bagdad had been the capitals of major empires. In each of those cases either the empire ceased to exist or the capital was moved. In the case of Constantinople the decline in population was associated with a decline in the empire of which it was the capital. In all these cases, the city's reason for being went away, and the population seemed to follow it.

However, the possibilities for conscious redesign of a community are probably limited. Haas et al. (1977) argues that "despite the best efforts to shape the character of the reconstructed city, fundamental change is unlikely." "Overambitious plans to accomplish … goals [such as improved efficiency, equity, amenity] tend to be counterproductive." They argue that private actors want a prompt return to normal. The longer the government delays, the more likely it is that people will rebuild, with or without government support.

Rubin (1985) supports these conclusions. She repeats the observation that if government delays too long in planning for a post-disaster future, the more likely it is that private action will make such planning moot. Furthermore she notes that consensus on reshaping a community is not always easy to obtain. In her case study of the 1983 Coalinga Earthquake she notes that Coalinga had trouble precisely because there were two competing visions and no consensus.

2.3.3 Population Recovery

As discussed above, community populations almost always recover to their previous levels. For the most part, exceptions are limited to cities that are already in long-term decline. A particularly dramatic example of long-term recovery is the city of Warsaw, Poland. During World War II, Warsaw was almost completely destroyed, and had lost more than 80 % of its pre-war population. But within 15 years its population returned to pre-war levels.

2.3.4 Physical Infrastructure

Studies of time to rebuild commercial and industrial infrastructure do not seem to exist. Any idea on the rebuilding of commercial and industrial infrastructure must be inferred from data on the restoration of business services. Similarly, little data exists on recovery of inventory losses.

There are a number of models regarding recovery of lifeline infrastructures (like power, gas, water, transportation). There are very few regarding longer term lifeline recovery. Models of response and recovery include Isumi et al. (1985), Kozin and Zhau (1990), and Liu et al. (2007).

Business recovery is heavily dependent on restoration of lifelines. For example, Tierney (1997a), in her survey of businesses following the Northridge earthquake, found that 58 % of business closures were in part due to lack of electricity, 56 % due to the inability of employees to get to work, and 50 % due to lack of telephone service (multiple reasons were cited by most businesses). Gordon et al. (1998) concluded that 27 % of business interruptions were due (in part) to transportation interruptions, with job 'losses' of more than 15,700 person-years.

Following the Kobe earthquake, electric power took nearly a week to restore, telecommunications two weeks, water and natural gas about two and a half months, railway seven months, and highway transportation 21 months (Chang 2010). Electrical power was restored within 28 hours following the Northridge earthquake in all but two service areas (Rose and Lim 2002). Median restoration times for water and natural gas were about two days; restoring sewage services took an additional day for affected businesses (Tierney 1997a).

Quarantelli (1982) argued that there are four stages involved in housing recovery:

The first stage is emergency shelter, which consists of unplanned and spontaneously sought locations that are intended only to provide protection from the elements. This period typically lasts only a few hours, and falls in the Response period.

The next step is temporary shelter, which includes food preparation and sleeping facilities that usually are sought from friends and relatives or are found in commercial lodging, although 'mass care' facilities in school gymnasiums or church auditoriums are acceptable as a last resort. As a rule this lasts only for a few days, and also falls in the Response period. This was discussed briefly there.

The third step is temporary housing, which allows victims to reestablish household routines in nonpreferred locations or structures.

The last step is permanent housing, which reestablishes household routines in preferred locations and structures. In some cases temporary housing ends up becoming permanent housing (Bolin and Stanford 1991), usually against the intentions of the agencies providing the temporary housing.

Bowden, et al. (1977) observed a number of regularities in rebuilding. Their main example is the 1906 San Francisco Earthquake. The wealthy reestablished housing first. The poor were several years in finding new houses, and many who lived there during the disaster left. Initially, with 50 % of the housing stock destroyed, and the population not seriously depleted, housing prices were very high. As time went on, and the building stock was restored, prices fell. They found similar price effects in the Rapid City, SD flood of 1972. Bolin and Stanford (1991) found that earthquakes tended remove the bottom tier of housing from the market. The bottom tier was typically either destroyed or had to be upgraded to such an extent that the poor were priced out of the housing market.

In San Francisco, the city expanded and reduced its density. The Central Business District grew in physical size, and residences moved further out. Sorting became much stronger. So, rich and poor neighborhoods that had been somewhat intermixed before became strongly sorted. Ethnic enclaves became much sharper. The business district became much more sorted with the financial district cleanly separated from the garment district, from the theater district etc.

Other researchers have found similar results. Chang (2010) found after the Kobe (1995) earthquake that there was increased suburbanization. Friesma (1979) found the same result in Topeka, KS after flooding there.

2.3.5 Economy

Friesma et al. (1979) suggest that disasters can be good for the economy. The basic argument is that it results in replacement of capital stock with newer and better, which in turn makes the community more competitive. The argument is difficult to support. If companies and regions would become more competitive by upgrading their capital, they have a powerful incentive to do so even absent a disaster.

What the disaster may do is get someone else to fund capital improvement. Money transfers to a disaster-stricken area can be quite large. With aid, the *region* may end up economically better off than it was before, but the nation is still worse off. The money that went to fund reconstruction still has to be financed either out of consumption or by deferring or foregoing some other investment.

Even with aid, however, it seems unlikely that regions are often better off for disasters. Chang (1983) evaluated the effect of Hurricane Frederick on the finances of the City of Mobile, AL. He found that the impact of the hurricane was net negative. Chang (2010), in her analysis of the recovery of Kobe, found that Kobe never fully recovered. There was an initial construction boom "followed by settlement at a 'new normality' that was roughly ten per cent lower than the pre-disaster normal." Guimaraes et al. (1993) estimated the economic impact of Hurricane Hugo on the Charleston, SC economy and found a similar construction boom followed by a construction lull. Apparently much of the repair, maintenance, and improvement that would have normally been spread out over several years was compressed into the time period immediately after the hurricane. A similar effect was visible in forestry. The hurricane did enormous damage to the Frances Marion National Forest. The forestry industry went through a short-term boom after the hurricane due to a massive salvage effort, but the boom was followed by a bust due to the dearth of harvestable trees.

Bowden, et al. (1977) in their analysis of recovery from the 1906 San Francisco Earthquake found that trends in terms of business types accelerated. So manufacturing was already moving out of San Francisco, and that trend accelerated. Chang (2010) found a similar result for Kobe after the 1995 Earthquake. In Japan, over the long-term the proportion of small businesses had been declining. In Kobe that trend was greatly accelerated by the earthquake. Webb et al. (2000) found that businesses in declining industries tended to do worse in disasters.

Most studies find that small businesses are hardest hit by disasters (Tierney 1997a, 1997b; Webb et al. 2000). These results do not attempt to take into account normal business turnover. Even without disasters small businesses have a much higher turnover than large businesses. It is not clear from these studies how much of the small business losses in comparison to large business are attributable to their higher vulnerability and how much to normal turnover.

Business preparedness had no measurable impact on business recovery. Webb et al. discuss some reasons why preparedness does not seem to impact recovery. 'Preparedness' is usually operationally defined as having taken one or more of a specified set of measures. The measures are determined by the experimenter and are usually measures that are typically recommended by experts. Webb et al. argued that the measures that operationally define preparedness are largely aimed at improving emergency response and have little impact on recovery.

Construction does better after disasters than do other businesses and typically financial and real-estate firms do worse. Businesses as a rule effectively self-insure against disasters. Very few in any study have insurance against losses.

Webb et al. found that aid does not have any positive impact on recovery, while Dahlhamer and Tierney (1998) found that aid was actually negatively correlated with recovery. The most likely explanation is that firms don't apply for aid unless they are in bad financial shape. In addition, business aid nearly always takes the form of loans. The additional leverage could make business survivability worse. If so, then only businesses on the brink of bankruptcy will be willing to apply for such aid.

In all of these studies it is not clear to what extent survivor bias[9] plays a role in the conclusions. Most of the studies of contributors to business recovery (including Chang and Falit-Baiamonte 2002; Dahlhamer and Tierney 1998; Tierney 1997a, 1997b; Webb et al. 2000) pass out surveys to businesses identified in the affected area. Efforts are made to target businesses that had been in the area before the disaster struck, but it is not clear how effective those efforts are nor to what extent response is correlated with survival.

2.3.6 Social Networks, Government Services and the Environment

There is very little in the literature on recovery of social networks, government services or the environment from disasters.

Disasters in many ways produce behavior that is strongly prosocial. People help out their neighbors, crime and looting are rare, abundant aid comes in from outside to the affected area (Waugh and Tierney 2007). However, there are also sources of conflict as well. Much of the conflict is over distribution and use of aid (Berke et al. 2008), the shape of the post-disaster community (Rubin 1985), or temporary housing (Bolin and Stanford 1991).

Temporary housing is typically disruptive of social networks. Specifically, it is generally not designed in such a way as to respect preexisting social networks and so ends up breaking them up (Bowden et al. 1977).

2.3.7 Simulating Recovery

Miles and Chang (2003) develop a numerical model simulating recovery. Their model includes modules for households, businesses and lifeline systems. The households, businesses and lifeline systems are organized into neighborhoods and communities. There are minimal markets for labor and consumer goods. The model allows both businesses and households to leave the region.

The major factors in the behavior of the model end up being lifelines and household income. If lifelines go down, negative impacts on the model are large and all the households may end up leaving the region.

They attempted to simulate the aftereffects of the Kobe earthquake using the model. The model did not have a very good fit. In at least one simulation, using parameters that seemed reasonable, all the simulated population of Kobe left.

Their model does not account well for the economics of such a situation. Markets for labor and goods are minimal. And the decision criterion for households leaving is somewhat *ad hoc*. Their

---

[9] "Survivor Bias" is the bias that results when your data comes primarily from survivors. In this case, if businesses that survive are more likely to be interviewed (or to respond to surveys) then the significance of any factor that correlates with survival will be misidentified.

model would provide one interesting place to start in simulating disaster recovery, but more work is needed.

Glaeser and Gottlieb (2009) discuss location choice in the context of urban economics. Urban economics assumes that people choose where they work and live so as to maximize their utility. This would apply in particular to the decision as to whether to leave an area after a disaster. To illustrate this idea, suppose there are only two cities in the world, which we will dub New Orleans and Memphis. Prices, wages, and local amenities are different in each city. Amenities are assumed to be public goods (and hence free). The choice sets of consumption goods are identical between the two cities, although, as mentioned, their prices may be very different. Assume a person lives in New Orleans. Then the utility of living in New Orleans is:

$$\max \quad ,$$

Subject to the budget constraint:

Where:
a_{NO} = the amenities New Orleans offers (which could be positive or negative)
c = the consumption vector,
p_{NO} = the price vector in New Orleans,
w_{NO} = wage in New Orleans,
L = labor
r = rate of return on capital (assumed to be identical in both cities)
K = savings

Similarly for Memphis we have:

$$\max \quad ,$$

Subject to the budget constraint:

The only new variable here is m, which represents moving costs.

Before the disaster strikes, the equilibrium assumption implies that $U_{NO} \geq U_M$ for residents of New Orleans. The disaster would tend to cause amenity (a_{NO}) and wage (w_{NO}) to decrease, and price level (p_{NO}) to increase. All of these serve to increase the likelihood of ex-migration. In addition, K will likely decrease for residents of New Orleans, however its effect on ex-migration is ambiguous. Price level would increase due to the scarcity of goods. In particular if a significant fraction of the housing stock is destroyed, home prices and rents will increase significantly, increasing cost of living.

This simple model suggests the most important factors in preventing ex-migration are a rapid restoration of jobs and housing. Depending on the importance of amenity in people's preferences, rapid restoration of amenity may also be important. Since there are fixed costs to moving, it seems likely that such changes would have to be large or of long duration before significant migration effects appeared.

To see the impact of capital loss (e.g., destruction of an owner-occupied dwelling), assume there is only one consumption good, and utility is based solely on consumption (that is, amenity is irrelevant). Then the equilibrium assumption implies that consumption in New Orleans is greater than consumption in Memphis. That implies that:

$$\frac{1}{\ } - \frac{1}{\ }$$

Reorganization gives:

$$\frac{1}{\ } - \frac{1}{\ } \quad 0$$

Then

$$0$$

A reduction in capital increases the likelihood of ex-migration if   . Otherwise, it reduces it. In other words, reductions in owner-capital favor high-priced regions and disfavor low-priced ones.

Some researchers suggest that in addition to the factors discussed above a home serves as a liquidity constraint that limits the ability of people to move. In such a case, destruction of the house could serve to lift the liquidity constraint. Alternatively, by destroying assets (and not debts) the disaster could inhibit migration by making it more difficult to finance moving.

A theoretical macroeconomic model of recovery can also be constructed. Assume there is a single representative consumer, with access to a Cobb-Douglass production technology. Assume Constant Elasticity of Substitution preferences. Then the representative consumer seeks to maximize present expected utility of consumption:

$$\sum^{\infty} \frac{1}{1}$$

Subject to the budget constraint:

Where

C = Consumption

K = Capital

α = Cobb-Douglass production parameter; related to capital intensity

δ = depreciation rate of capital

ρ = discount rate

σ = preferences parameter; related to the intertemporal elasticity of substitution

x = exogenous economic growth rate.

Mathematical development is in the mathematical appendix.

Assume that the pre-disaster city is on the balanced growth path. Then a disaster is a reduction of capital below the balanced growth path. Recovery time is modeled as the amount of time required to return approximately to the balanced growth path.[10] Estimated time to recovery is shown in Figure 5. This is best thought of as time required to return to the state that would have obtained had the disaster not occurred. Figure 6 shows the amount of time required to return to the levels of capital and consumption that obtained before the disaster. The values are different because the productive capacity of the economy is growing even as the region is recovering from the disaster.

Figure 5: Percent recovery versus time (in years) for different levels of destruction in a disaster

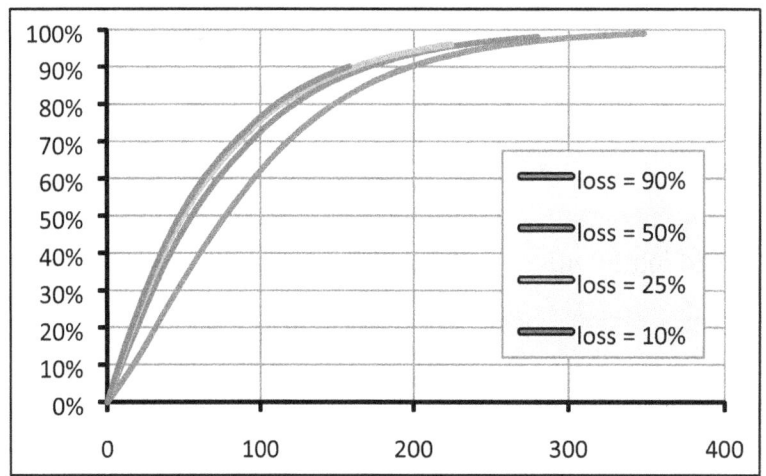

The model makes two unrealistic assumptions that impact the results. First, the model assumes that the region impacted by the disaster is a closed economy. The region depends entirely on its

---

[10] Strictly speaking, in this model, capital levels approach the balanced growth path asymptotically. Approximate return is defined as arriving within a predetermined distance of the balanced growth path.

own resources for recovery. Second, there is no construction sector, so scarcity of construction cannot impact time to recovery. The first assumption significantly extends the time to recovery, while the second assumption shortens it. On net, this model probably overstates significantly time to recovery. A more complete model, correcting for the limitations mentioned above, would provide a more realistic estimate of recovery time.

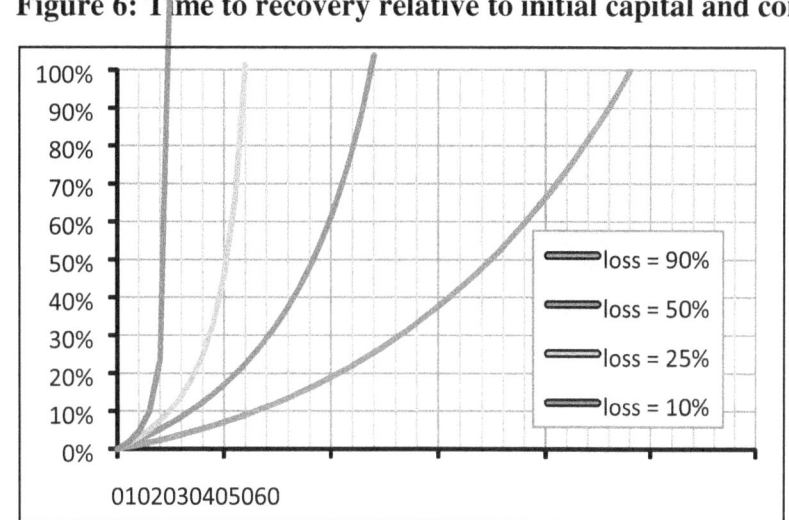

Figure 6: Time to recovery relative to initial capital and consumption

2.3.8 Ways to Improve Recovery

As with response, it is difficult to identify clear lessons for improving recovery. Preparedness does not correlate well to recovery (Tierney et al. 2001; Webb et al. 2000). Aid seems to influence individual recovery, but not always for the better (Bolin and Stanford 1991). On the other hand, aid seems to be at best uncorrelated to business recovery (Webb et al. 2000).

Bolin and Stanford suggest that aid aimed primarily at providing permanent housing can enhance individual and family recovery, while aid aimed primarily at providing temporary housing can impede it.

Rubin (1985) suggested that three things mattered to recovery: Leadership, the Ability to Act, and Knowledge. She argued that recovery was enhanced when the community settled promptly on a vision for the rebuilt community (Leadership), effectively interfaced with aid agencies while maintaining local control over the spending of resources (Ability to Act), and knew how to obtain aid and resources (Knowledge).

Berke et al. (2008) argue that good aid programs are responsive to household needs and are implemented by strong organizations capable of achieving program goals. They argue that many aid programs fail because the aid does not get to the groups that most need it. Such groups often do not have the information to seek out aid that is available or the political weight to convince aid donors to provide them with aid. Aid also often fails because the agencies providing aid do not have the local knowledge to tailor aid to the needs of the community.

They argue that good aid has four characteristics: (1) Goodness of fit to local needs and opportunities; (2) reliance on local resources in planning, distribution and implementation; (3) local knowledge about how to obtain disaster assistance; and (4) flexibility and willingness to learn in process.

2.4 Preparedness

The NFPA (2007) defines preparedness as "Activities, tasks, programs, and systems developed and implemented prior to an emergency that are used to support the prevention of, mitigation of, response to, and recovery from emergencies." *Preparedness* is defined in this paper as measures taken in advance of a disaster with the objective of improving response and recovery. It specifically includes warning and evacuation.

Preparedness usually is operationally defined as fulfilling a checklist pre-selected by the researcher (e.g., Godschalk et al. 1999 who operationally define preparedness as meeting the requirements of the Stafford Act).

As discussed above, preparedness by businesses does not correlate with loss or recovery. Similar results hold for communities. Tierney et al. (2001) observe that planning does not correlate well with effective response or recovery. Wenger et al. (1986) noted in their case studies that planning did have some positive effect, but many of the jurisdictions with plans did not handle disaster response well. They noted that "more problems and difficulties arose in those communities which had no planning or very limited planning activities. Second, while moderate extensive planning activities tended to facilitate an effective response they did not guarantee it." In their case studies the key to effective response really was extensive previous experience.

Why does planning have such ambiguous effect? There are at least three possibilities.

Possibility 1: In practice there are limits to what can be planned for. Resource and human constraints limit us. Regarding resource constraints, nature has more resources than we do and will occasionally overwhelm even the best of plans. Regarding human constraints, there are limits to how far people will depart from their normal behavior even in disasters.

Possibility 2: Information constraints. It is not clear what a plan should include. The check-lists may be made up of items that have little to do with resilience and recovery. Unfortunately, learning is hard.

Possibility 3: Checklists are necessarily made up of things that can be easily verified. It may well be that the things that actually contribute to resilience and recovery are not easily verified. In this case, the checklist is essentially noise covering up the signal of how well prepared a jurisdiction actually is.

All three probably play a role, although their relative contribution is not known.

2.4.1 Warning Systems

Warning and evacuation systems are generally regarded by researchers as one of the most effective ways to prevent death and injury in disasters.

Mileti and Sorensen (1990) did a meta-analysis of studies of warning systems and suggested several lessons learned from the literature. First, they repeated the lesson that people don't panic when warned of an impending disaster. Their reaction to warnings is generally to seek more information; people generally want more information than they get. They also tend to want information from multiple sources.

People do not simply do what they are told—they make their own decisions about what to do. However, most people can be persuaded. Again, that requires providing people with the information they need to make their decision.

They suggest that false alarms do not necessarily diminish the effectiveness of future warnings if the reasons for the false alarm are explained. They point out that, at the time they wrote their paper, about 70 % of hurricane evacuation warnings were false alarms. It seems likely that the false-alarm issue is more complicated than they suggest. A false alarm is (at best) evidence of ignorance, and reacting to warnings is costly. If false alarms represent a high enough percentage of warnings then the cost of reacting to a warning will exceed the expected benefits. At that point, people will quit paying attention to them. However, people react to warnings by seeking more information. Since the cost of seeking more information is relatively low, people will tolerate a relatively high percentage of false alarms before they begin to ignore them.

They suggest that sirens aren't much use as warning tools. People do not know what the siren is warning against. The normal response to warning sirens is to seek more information about what the hazard is. However, they note that very few warning systems are capable of providing short-term warnings (under three hours).

The standard progression for how people react to warnings is Hearing, Understanding, Believing, Personalizing (is it relevant to me or not), Deciding and Responding, and Confirming (going back to check the current status).

2.4.2 Evacuation

The research on evacuation starts with two standard observations. First, when evacuations are ordered not everyone leaves (Tierney et al. 2001). Second, some people leave even though they have not been ordered to evacuate. The latter phenomenon is called an 'evacuation shadow.'

Fothergill (1998) reports that women are more likely to evacuate than men. Women generally have better social networks than men and so are more likely to hear about an impending hazard.

Women are also more risk averse, and so are more likely to evacuate once they have heard about the hazard. Family behavior falls somewhere in between, with men often agreeing to evacuate at their wives' insistence.

Some people stay because they don't hear, believe, or personalize the warning. Tierney et al. note for example that some ethnic groups are routinely distrustful of authority and may not believe warnings when given. Some people cannot leave: for example some of the poor do not have transportation, and some people are disabled. In the U.S., the assumption has been that people have their own transportation and can manage their own evacuation. Hurricane Katrina showed that that was not always true.

Figure 7: Conceptual degree of evacuation versus risk

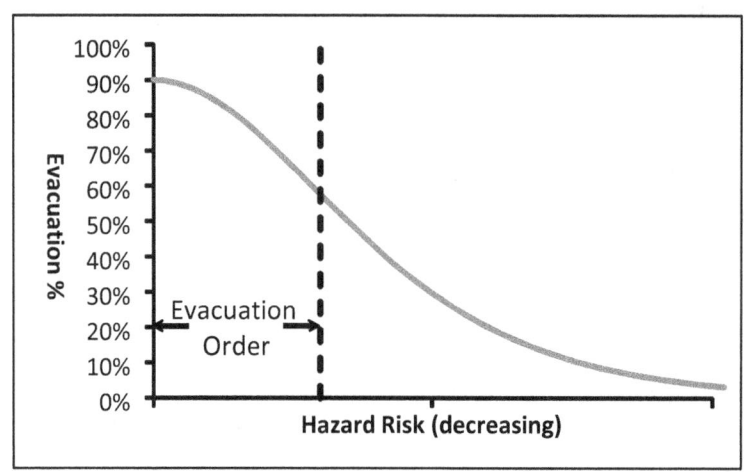

Evacuation is risky, difficult, and expensive. When southeast Texas evacuated ahead of Hurricane Rita a number of people died as a result of the evacuation (e.g., Belli and Falkenberg 2005).[11] Spending several days somewhere other than in one's own home carries both financial and hedonic costs.

In general we would expect the decision to evacuate to be based on a balancing of the risks associated with remaining against the costs and risks of evacuation. A person's degree of risk aversion will mediate that decision. Risks of staying, risks of evacuating, costs of evacuating, ability to delay decision-making and seek better information, and risk tolerance will be different for different people. It is therefore unsurprising that evacuation behavior is heterogeneous. Conceptually, as shown in Figure 7, we would expect degree of evacuation to decrease as the risk from the hazard decreases. An evacuation order (hopefully) provides additional information to people in making their decisions, but it seems unlikely that they will completely delegate such

---

[11] The author personally knows someone whose elderly mother died as a result of being evacuated ahead of Hurricane Rita.

a decision to distant authorities who do not know their individual circumstances and cannot take them into account.

2.5 Mitigation

The NFPA (2007) defines *mitigation* as those "activities taken to reduce the severity or consequences of an emergency." In this paper *mitigation* will be defined as measures taken before a disaster strikes with the objective of improving disaster resistance. We explicitly exclude preparedness. However much of the discussion below applies to preparedness as well.

A number of factors contribute to the determination of optimal mitigation (see Figure 8). An understanding of individual behavior is needed because it impacts and limits our ability to implement specific mitigation measures. In order to determine optimal mitigation we also need to know the type, frequency, and severity of the hazards; how vulnerable our communities are, and where those vulnerabilities lie. Those enable us to estimate probabilistic damages, which in turn we can use to estimate optimal mitigation and insurance. Each of these topics is investigated below.

Figure 8: Elements of mitigation, with dependencies

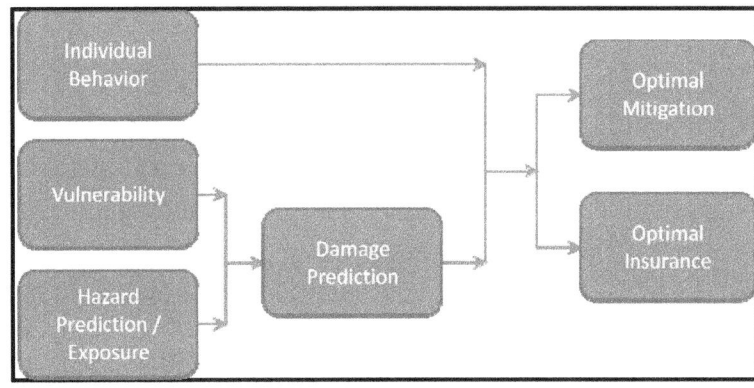

2.5.1 Individual Behavior

It is generally believed that people under-invest in insurance and mitigation compared to what is optimal (e.g., H. Kunreuther, Meyer, and Van den Bulte 2004). A number of research papers have explored reasons why people do not optimally invest in mitigation and ways to encourage optimal investment.

Kunreuther (2001) suggests that the reasons people fail to invest optimally in mitigation include bounded rationality and moral hazard induced by the expectation of disaster aid. Bounded rationality reduces investment because boundedly rational people tend to underestimate the risks of low-probability events.[12] Moral hazard reduces investment because when people expect a portion of their losses to be paid for by disaster aid (or insurance) they have less of an incentive to invest in loss avoidance. Similarly, Muermann and Kunreuther (2008) analyze a theoretical model in which people fail to invest optimally in mitigation due to externalities.

Alternatively, Braun and Muermann (2004) suggest that if regret is a factor people will be more likely to buy insurance, but less likely to buy full insurance than if they were expected utility maximizers. Regret theory (Loomes and Sugden 1982) suggests that people's enjoyment of a specific outcome is diminished if there is a better outcome that they could have enjoyed had their choices been different. People for whom regret is important are less likely to invest in full insurance because if there is no disaster they will regret having bought insurance. On the other hand, they are less likely to buy no insurance because if there is a disaster they will regret not having bought insurance.

Lindell and Perry (2000) and Lindell and Prater (2000) find that the best predictors of adoption of mitigation are risk perception and personalization of risk. Demographic variables (race, sex, age, etc.) fare poorly. Lindell and Prater (2000) note that hazard experience matters to risk perception. They also find that women are more risk averse, but do not do more mitigation.

2.5.2 Vulnerability

Vulnerability differs across people. Poverty, lack of education, disability, and minority race are all correlated to vulnerability (Tierney 2006). However, since the factors listed are highly correlated, it is not clear how much they independently contribute to vulnerability. Tierney observes that one of the contributors to their vulnerability is that disaster preparedness and assistance programs are not all that well tailored for the poor and disadvantaged. This includes, among other things, that wealthier victims are also better able to search for and find assistance.

This does not hold universally for all disasters. Gordon et al. (2002) attempt to estimate the distributional effects of an earthquake on the Elysian Park fault in Los Angeles. In that case they find that the well-off suffer a disproportionate share of the damage. They suggest that the reason is that information on spatial distribution of earthquake hazards may be harder to come by, so the relatively well-off may not be able to avoid the risk.

Cutter et al. (2003) estimate a Social Vulnerability Index for counties in the United States. They took 42 variables, drawn from 1990 census data, which are associated with increased individual vulnerability. They used a principle-components analysis to reduce those variables to 11. The 11 variables were then added together (and rescaled) to form the Social Vulnerability Index (SoVI™). They tested the index against Presidential disaster declarations to estimate its explanatory power. The index was negatively correlated with Presidential disaster declarations.

---

[12] The standard observation is that people either under- or over-estimate the risks of low-probability events.

In a separate paper, Cutter (2008) applies the framework to estimate the spatial distribution of vulnerability in New Jersey.

The negative correlation with Presidential disaster declarations is not as problematic as it seems. Disaster declarations are known to have political components. Downton and Pielke (2001) found that disaster declarations for floods are more likely in election years, even after taking weather into account. Garrett and Sobel (2003) found that counties are more likely to receive disaster declarations if they are in a 'battleground' state in a Presidential election year or if they are represented by a member of congress who sits on one of the FEMA oversight committees. The same factors that contribute to vulnerability also tend to reduce political weight.

Nevertheless, the SoVI has serious flaws. Income at the county level, which is one of their main variables, is problematic. Prices vary dramatically between counties, so real income levels have much less variation than any measure of simple wealth (Glaeser and Gottlieb 2009). Second, they do not attempt to correlate their vulnerability factors to actual harm done in disasters. As a result, it is not at all clear whether some of the factors are overweighted or underweighted. Estimating against actual harm would be difficult because major hazards are rare and spatially random, but failure to do so reduces the predictive value of the index.

Using similar methods Cutter (2008) develops an Infrastructure Vulnerability Index and an Environmental Vulnerability Index for the State of New Jersey. Borden et al. (2007) develop a Built-Environment Vulnerability Index and a Hazard Vulnerability Index for counties containing 132 metropolitan areas in the United States.

Factors that are believed to contribute specifically to community vulnerability are economic and social diversity, social cohesion, and communication among members of the community (Norris et al. 2008; Tierney et al. 2001).

Business vulnerability is strongly associated with being small and being in retail (among others, Chang and Falit-Baiamonte 2002). However, as noted earlier, these results do not take into account normal turnover. These same types of businesses tend to have high turnover even without disasters.

2.5.3 Hazard Models and Damage Prediction

In order to determine the optimal mitigation, we need to have an estimate of how likely disasters are and how much damage they will do. Grossi et al. (2005) discuss the basic approach used to estimate this. A number of models and services provide probabilistic estimates of damage as described in Grossi et al., including HAZUS-MH provided by FEMA, the EQECAT catastrophe modeling system published by EQECAT, Inc., and the catastrophe modeling and consulting services of RMS Inc. All these systems include estimates of hazard probability and estimates of damage resulting from a given hazard. Damage estimates include estimates of lives lost, damage to the built environment, and indirect economic losses. The private models are aimed primarily at insurance companies and are intended to estimate insured losses.

Estimates of damage to the built environment are typically based on fragility curves for different classes of structures, combined with location-coded inventories of structures (for example Kappos and Dimitrakopoulos 2008).

Estimates of indirect losses and losses to the economy are based on simplified regional macroeconomic models. For example, HAZUS-MH has the regional accounting system from IMPLAN at the core of its indirect economic cost module (Rose 2002). This has been a very active part of the literature in the last 15 years, with a number of techniques developed and substantial improvements and refinements to the approaches. Most of these will be discussed in more detail below.

A couple of examples from the literature where estimates of losses were made for possible future disasters are Ellson et al. (1984) who estimate the impacts of an earthquake in the Charleston, SC area, and Rose et al. (1997) who estimate indirect losses in Memphis, TN due to electrical outages caused by an earthquake along the New Madrid fault system.

Bernknopf et al. (2001) proposed a simplified approach to damage estimation for certain types of earthquake damage. They demonstrate their technique by performing an assessment of the risks of earthquake-induced lateral-spread ground failure in Watsonville, CA. They use property tax records to provide a reconnaissance level estimate of building inventory and type. They use soil and geologic maps and topographic maps to estimate the regional distribution of soil types. Using a ground-shaking model they estimate the risk of failure for each soil type. That combined with the property-tax inventory allows them to estimate damages from an event.

Complicating the problem of hazard estimation is that disasters often have other follow-on problems that are caused by the initial disaster. The standard example is fire following earthquake. In the 1906 San Francisco Earthquake the majority of buildings destroyed were destroyed by fire rather than the earthquake itself. Many modern hazard models include this as part of the model.

2.5.4 Optimal Mitigation

There are a number of standards relating to mitigation, including ASTM Standard E 2506 (2006) and NFPA Standard 1600 (2007).

In general the approach to determining optimal mitigation consists of (ASTM 2006):

- Establishing risk mitigation objectives and constraints;
- Identifying and selecting candidate combinations of risk mitigation strategies;

- Computing measures of economic performance for each candidate combination, taking into account uncertainty and risk;

- Analyzing results and recommending the most cost-effective combination of risk mitigation strategies.

Risk mitigation objectives must balance the benefits of mitigation (fewer deaths and injuries, less value of buildings destroyed, less business interruption costs, etc.) against the costs. In many cases that is simply a comparison of dollars to dollars.[13] However, when the objective involves avoiding non-monetary damages (like deaths or injuries avoided, environmental damage avoided, etc.) the problem becomes more difficult. In order to make that decision, a monetary value for lives saved (or environmental damage avoided, etc.) must be explicitly or implicitly arrived at.

Approaches to estimating such values include attempts to determine how much value people themselves place on avoiding death (or injury, or environmental damage, etc.). Methods to do this include contingent valuation surveys where people are essentially asked how much they value the death avoided using a carefully-drawn questionnaire (Carson and Hanemann 2005), and econometric studies that estimate how much people give up to avoid premature death in their daily lives. A relevant example of the former is Viscusi (2009), who estimated willingness to pay to avoid deaths due to natural disasters and terrorism as compared to traffic deaths. He found that natural disaster deaths are worth half of traffic deaths. Terrorism deaths are worth about the same as traffic deaths.

A second approach to valuing deaths effectively compares lives saved to lives forfeit. Life expectancy is a normal good—the greater people's income the more they are willing to spend to increase life expectancy. Spending money to avoid one hazard reduces the amount people have available to spend to reduce other hazards. If enough money is spent to save lives from (say) natural disasters, then the reductions in spending on other hazards will result in a life forfeit from other causes. Lutter et al. (1999) investigated this and estimated the value of spending per life saved at which lives lost due to income effects equal lives gained from the spending program at about $15 million per life saved. Programs which cost more than that per life saved effectively cost more lives than they save.

Virtually all mitigation measures discussed in the literature fall into four categories: improved building and construction practices, avoiding hazardous areas, protective structures (both natural and man-made), and warning and evacuation (e.g., Godschalk et al. 1999; Mileti 1999). Warning and evacuation was already discussed under Preparedness.

---

[13] Note, however, that the comparison is typically of guaranteed dollars today (for the mitigation) to uncertain dollars sometime in the future (for the avoided damages). That requires decisions regarding degree of risk aversion and time preference in order to make those comparisons.

An example of improve building and construction practices is the report by the San Francisco Planning & Urban Research Association (SPUR). They produced a report (SPUR 2009) that proposed a series of improvements to building codes and construction practices for San Francisco. The improvements were intended to improve the survivability of the built infrastructure of the city in a major earthquake. They developed a set of performance standards for buildings and lifeline infrastructure and a series of recommendations regarding improvements to building codes and construction practices intended to help the city meet those standards.

Protective structures would include man-made structures like dams and levees, but would also include natural barriers like protective coastal dunes (Godschalk et al. 1999) and wetlands (Brody et al. 2008).

Avoiding hazardous areas is the mitigation measure strongly preferred by the sociology literature (essentially, "don't build your house where there is driftwood in the trees"). The most frequently mentioned examples are the removal of housing in flood plains and on barrier islands in hurricane-prone regions.

A number of papers have been written on identifying optimal mitigation measures. Most use a damage-estimation model, run the model several times using different candidate mitigation policies, compare them using some predetermined preference relation, and identify the best from the set.

The series of papers by Dodo et al. (2005, 2007) and Xu et al. (2007) use HAZUS as their damage estimation model. They evaluate building upgrade policies. The number of policies evaluated in their examples numbers in the thousands. They use linear programming techniques to estimate the optimal policy, and develop numerical methods for efficient finding of a solution.

Chapman and Rushing (2008) and FEMA (2005) are implementations of the ASTM standard, and are intended to aid in determining optimal mitigation. FEMA (2005) is primarily focused on hazards due to terrorism. It provides detailed guidance regarding the first two elements of the standard, but provides relatively little guidance regarding computing measures of economic performance. In contrast Chapman and Rushing (2008) largely assume that the first two steps have been completed and provide a detailed protocol and software tool for estimating economic performance.[14] They also allow the separate explicit modeling of non economic damages from hazards. Thomas and Chapman (2008) provide a detailed guide and annotated bibliography of sources useful in carrying out this evaluation.

2.5.5 Optimal Insurance

In general, insurance does not reduce damage in disasters. Rather, insurance is a means of redistributing the losses. In that sense, disaster aid programs serve as a form of insurance. It is

---

[14] They allow the user to determine the rate of time preference, but implicitly assume that their users are risk neutral.

generally believed that people exposed to catastrophic risk under-insure (Howard Kunreuther and Michel-Kerjan 2009).

Insurance in disasters presents special problems that do not apply to insurance in general. From an insurance perspective disasters are characterized by having highly correlated losses. For example, with homeowners insurance, normally if one house burns down it is highly likely that all the other houses in the area are still intact. In a normal housing insurance market, houses burn down all the time, but they are a small and predictable proportion of the entire population of houses.

In disasters, however, many houses are destroyed at the same time. That means that an insurance company insuring against disasters must maintain much higher reserves to stay solvent and still pay off on its policies. One of the strategies for dealing with this is reinsurance. Ermoliev et al. (2000) and Amendola et al. (2000) both analyze the optimal reinsurance problem for insurance companies exposed to catastrophic risk.

The U.S. GAO (2007) analyzes the catastrophe insurance market in the United States. The report responded to the impression that the catastrophe insurance market is failing. They suggest that a private insurance market where firms charge premiums that fully reflect actual risks, where there is broad participation by insurance firms, and where the market limits costs to taxpayers both before and after a disaster may be unobtainable, especially given the incentives by state insurance regulators to keep premiums below prices that reflect actual risk.

When liquidity constraints are binding, it is possible for insurance and disaster aid to reduce damages in disasters. Investment is often necessary to limit indirect effects from disasters. For example, a business may shut down due to the destruction of vital machinery in an earthquake. The ongoing losses due to business interruption may continue until the damaged or destroyed machinery is replaced. However, if liquidity constraints are binding, the investment may be delayed or foregone, thus allowing the loss to continue. It is unclear whether any theoretical or empirical research has been conducted into this.

2.6 Data and Measurement

Data generally fall into four types. Case studies, insurance claims, direct measurements, and survey methods.

Case studies are used extensively in the social science literature studying disasters. They have the advantage of providing a lot of detail about the community, preparedness (e.g., Wenger et al. 1986), mitigation measures (e.g., Brody et al. 2008), response and recovery (e.g., Comfort 1999; Haas et al. 1977). That detail can be used to evaluate the effectiveness of these measures. However, estimates of damage, vulnerability or resilience are usually developed by the methods discussed below. The estimates are also usually qualitative, at least in application. Case studies have so far been the best source of information regarding the effectiveness of various mitigation and preparedness measures, and on response and recovery. However, their usefulness is limited by the fact that such studies have very small sample sets. Furthermore different studies are not usually comparable, and so the ability to do meta-analyses across multiple studies is limited.

Insured losses for disasters are compiled by a number of companies including Swiss Re, Munich Re, and (in the United States) by Property Claims Services, a service of Verisk Analytics, Inc. FEMA, which offers flood insurance, also collects insured loss data and data on disaster assistance. Insured losses have the advantage of providing a numerical estimate of losses due to a disaster, and are provided by organizations whose incentives are to get accurate estimates of losses. However, they are limited to insured losses. Insurance coverage is not complete, and not all types of losses are covered by insurance. Therefore estimates of total loss are biased down and have limited coverage. Using insurance penetration estimates it is possible to estimate total losses to the built environment, but data are too thin to estimate losses to other things we want to protect.

Direct measurements of losses have generally been collected by local agencies or the National Weather Service, and compiled by various government agencies (Cutter, Gall, and Christopher Emrich 2008), including FEMA (disaster declarations), NOAA (weather-related disasters), and the National Geophysical Data Center (geologic disasters). Damage estimates have often been based on 'windshield counts' (Downton, Miller, and Pielke 2005), and until 1995 were usually reported only to the nearest order of magnitude.[15] More recently other methodologies including remote sensing (e.g., Adams et al. 2004; RMS Inc. 2010) have begun to be used in the estimation of damages. Such damage estimates have largely focused on human losses and losses to the built environment. However, the breadth of coverage has increased over time.

Recently, damages from a small handful of disasters have been estimated using survey methods. The surveys were intended to estimate a certain class of damages, usually damage to the economy, and in particular business interruption costs. One example of such an estimate is Tierney (1997a) who used a survey to estimate business losses—and in particular business interruption losses—due to the Northridge Earthquake. Surveys are potentially subject to self-selection biases in terms of who responds to the surveys and to survivor biases. The degree of such biases is unknown.

Business interruption costs have begun to be estimated using a variety of econometric and regional macroeconomic models. The models, which are discussed in more detail below, attempt to estimate the economic impact of a disaster on the region compared to what conditions would have been like without the disaster. Using such models, estimates can be made as to business

---

[15] The National Weather Service, who collected and reported much of the disaster loss information, until 1995 reported loss estimates in preset categories. The categories were $50 000 - $500 000, $500 000 - $5 Million, $5 Million - $50 Million, etc.

interruption costs, and losses of production and jobs due to a disaster. However, such models are relatively hard to calibrate against actual losses.

Howe and Cochrane (1993) develop a set of guidelines for estimation of losses in disasters. Their guidelines are primarily aimed at estimating losses to the physical infrastructure, but they also discuss loss estimation techniques for business interruption, damages to historical monuments, and valuing losses of human capital and natural capital.

One of the complications in estimating damages is the fact that construction prices go up during recovery as construction capacity becomes relatively scarce compared to demand.

The Natural Research Council (1999) in its Framework for Loss Estimation, noted that "there is no system in either the public or private sector for consistently compiling information about their economic impacts." Nevertheless, a number of databases aggregate loss estimates from a number of sources to report loss data from disasters. These include SHELDUS, a product of the Hazards and Vulnerability Research Institute at the University of South Carolina, which compiles loss data for natural disasters in the U.S. since about 1960. The database www.flooddamagedata.com provides flood damage data for the U.S. between 1925 and 2000. The EM-DAT database is published by the Center for Research on the Epidemiology of Disasters, a division of the World Health Organization, and publishes data on losses from disasters world-wide.

All of these databases rely on many of the same data sources, and so cannot be considered independent of each other. SHELDUS consciously biases its estimates low. When the raw sources list a range of possible losses SHELDUS reports the low point in the range. The EM-DAT database routinely estimates losses higher than those drawn from other databases. It is also not particularly transparent, and it is not clear for any particular disaster what its source of information is (Gall, Borden, and Cutter 2009; Vranes and Pielke 2009). The database EM-DAT likely includes some indirect loss estimates, while SHELDUS and the flood damage database explicitly do not. They also reflect the underlying temporal bias of their sources, where more recent disasters tend to be better reported, and more classes of loss are included.

While, as discussed above, there are sources of data for human losses, losses to the built environment, and general economic losses, there are essentially no estimates for losses to Social Networks, Government Services and the Environment.

The National Research Council (1999) recommends that a single federal agency be placed in charge of compiling estimated losses—they recommend that the Bureau of Economic Analysis be given the task. They recommend that loss information should include who initially bears the loss (e.g., government, insurers, businesses, individuals, nongovernmental organizations), and the type of loss (e.g., property, agricultural products, human losses, cleanup and response costs, adjustment costs). They also recommend that efforts be made to estimate indirect losses such as (temporary) job losses and business interruption losses.

2.6.1 Recovery Measurement

There have been very few attempts in the literature to measure recovery. Recovery is usually measured in terms of time to full recovery. In attempting to measure recovery, there are two basic questions that must be answered. First, what are we measuring the recovery of? Most recovery measurements have measured recovery of population following major disasters. A few have measured recovery of the economy. The second question is what constitutes full recovery. It is generally argued that the proper comparison for recovery is not before versus after, but with versus without (for example Friesma et al. 1979). That is, a community has fully recovered when it has arrived at a state as good as what would have obtained had there been no disaster. However, as a practical matter, most measurements of recovery have considered the return to the *ex ante* state as full recovery. A few researchers suggest that the proper measure of 'full' recovery is attainment of some 'new normal' (Chang 2010).

Kates and Pijawka (1977) measured the recovery of a number of cities since 1200. They used population as the measure, and full recovery was defined as returning to the predisaster population. Their most severe disasters were products of war. In some of their disasters, population declines during the disaster exceeded 90 %. All of their cities eventually recovered. Cities with population declines of less than 70 % usually recovered in less than 50 years. Population declines of greater than 70 % tended to require centuries for recovery.

Chang (2010) develops a framework for the measurement of recovery and applies it to the 1995 Kobe Earthquake. She measures recovery of both population and of the economy. She considers all three definitions[16] of full recovery. Kobe returned to pre-disaster population levels in about ten years. It never returned to predisaster levels of economic activity, and it never returned to the former trends in either population or economic activity.

Chang develops an approach to operationalize attainment of a 'new normal.' She selects the greatest annual change from the few years preceding the disaster. When the annual rate of change following the disaster declines to within that level, a new normal is considered to have been achieved. She found that the economy attained a new normal within about 6 years, although some sectors (such as trade and services) attained a new normal level much more quickly.

Most researchers suggest that the proper comparison for recovery is with the disaster compared to without: that is, how does the region after a disaster differ from what would have obtained had there been no disaster. This, however, is not easy to operationalize. Both Chang (2010) and Friesma et al. (1979) operationalize it as return to trend. However, that is not the same thing. Other things are affecting a region at the same time as the disaster recovery is ongoing. For

---

[16] As described above, the definitions are:

1. When the community has arrived at a state as good as what would have obtained had there been no disaster.
2. When the community has arrived at a state as good as the *ex ante* state.
3. When some 'new normal' has been attained.

example, a recession hitting at the same time as the disaster would ensure that "with versus without" was not the same as return to previous trend. Chang (1983) evaluated the impact of Hurricane Frederic (in 1979) on municipal finances for Mobile, AL. His analysis was able to take into account the recession of the early 1980's for comparison to actual outcomes.

2.6.2 Individual and Family

Records of deaths due to disasters go back more than 100 years. As with all estimates of losses, older records are less accurate, and have poorer coverage. Records of numbers injured are less accurate than deaths and do not have the length of record. Quarantelli (1983) comments that in disasters, hospitals tend to skimp on record-keeping. So numbers of people treated and released, nature of injuries, and nature of treatment often is difficult or impossible to obtain. Records on population changes are usually available, although they are relatively low resolution.

2.6.3 Physical Infrastructure

Estimates of damage to physical infrastructure are available back to at least 1960. In some cases (flood damage for example) they are available back further than that. As always, the older the data is, the less reliable it is. Estimates on physical damage are almost always accurate to within an order of magnitude, and usually are more accurate than that (Downton and Pielke 2005).

2.6.4 Economy

There has been very little measurement of impacts on economic activity before 1990. Even since, measurement is highly incomplete. The insurance industry reports claims against its business-interruption insurance lines as part of insured losses. However, since the great majority of businesses effectively self-insure (e.g., Chang and Falit-Baiamonte 2002; Tierney 1997b), reported losses substantially underestimate total losses. In addition, there have been a handful of efforts to use surveys to estimate losses (e.g., Tierney 1997a, 1997b). But surveys have been inconsistent, less than comprehensive, and are potentially affected by self-selection issues and survivor bias.

Regional econometric models have been used to estimate losses in disasters, in particular in Hurricane Hugo (Guimaraes et al. 1993), Hurricane Andrew (West and Lenze 1994), the Northridge Earthquake (Rose and Lim 2002), and Hurricane Katrina (Hallegatte 2008). Business interruption losses have generally been less than 50 % of direct losses, although Hallegatte's simulations suggest that business interruption losses increase much more rapidly than direct losses. She suggests that if direct losses from hurricane Katrina doubled (to about $200 billion) indirect losses would exceed direct losses.

Most models used in loss estimation are either modified Input-Output models or are Computable General Equilibrium models. Input-Output models assume that input proportions for each good in the economy change slowly. In practice, the input proportions as usually assumed to be fixed for the duration of the model. So, a product for which electricity is one of its inputs will have its production reduced by half if the electricity delivered during the period of the model is reduced by half. This allows the effect of lifeline interruptions to be analyzed and to propagate the effects of business interruption through the economy. In practice, a number of modifications are made to the model to take into account limited input substitution and various forms of business resilience.

Computable General Equilibrium models assume that all markets are in equilibrium at all times. Inherently, they allow for much greater input substitution than Input-Output models, and they allow for the effects of lifeline interruptions to be allocated to the sectors where they have the least impact. Clearly, in the short term after a disaster, markets will not be in equilibrium. However as time goes on, the assumption becomes a closer approximation to reality. Rose (2004) recommends that "I-O models, with an adjustment for inventories, are probably better suited to recovery periods of less than one week, but CGE models are better suited to all other cases...."

Rarely have regional accounts been used in estimating losses from disasters. Guimaraes et al. (1993) compared regional accounts to a baseline regional econometric model of Charleston, SC to get estimates of losses from Hurricane Hugo. Hallegatte (2008) used regional accounts to validate her model of indirect losses from Hurricane Katrina.

Rose and Lim (2002) discuss the important distinction between stocks and flows in estimating losses to the economy. *Stocks* in this context would be losses that occur at a single point in time. For example, buildings, equipment, or inventories destroyed would be a stock. *Flows* refer to losses that occur over a period of time. For example, business interruption losses would represent a flow. Failure to properly account for them can easily lead to double-counting of losses. Rose and Lim argue that flows are the better and more natural loss to count.

Figure 9 shows a time line of production for an individual firm following a disaster.

Figure 9: Time line of production for an individual firm following a disaster

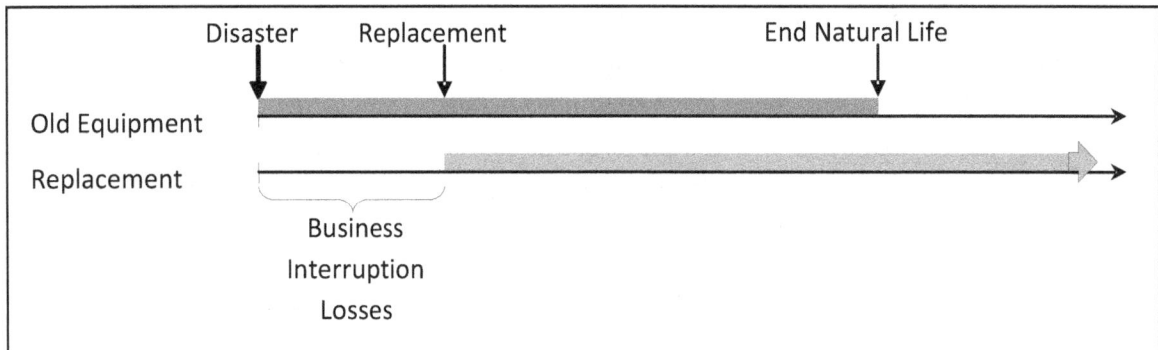

For a business like the one shown, total disaster losses to a first order of approximation would be:

$$L = D + R - V$$

Where:
L = Total Losses
D = Present expected value of production from equipment that was Destroyed.
R = Present value of net Replacement cost of equipment destroyed.
V = Present expected Value of production from new equipment.

Most estimates of damages have counted either D or R,[17] with estimates of replacement value (R) predominating. Since a firm will not buy a piece of equipment unless the net revenues it will generate exceed its cost, D will always overestimate the losses from a disaster.[18]

We can reorganize this in the following form:

$$L = B + R - \Delta V$$

Where:
B = Business Interruption Losses.
ΔV = Difference in the present expected value of production of the new equipment versus the old after production is restarted.

This casts the loss calculation in more traditional terms. Note that ΔV will usually be positive, so the sum of replacement costs and business interruption costs will usually overstate the losses to business.

As with economic accounts in general, there are three ways the business interruption losses and difference in values can be aggregated. By analogy with economic accounts, we could refer to these as the *expenditures* approach, the *income* approach, and the *value-added* approach.

The *expenditures* approach would count as business interruption costs all revenues for the production of final goods only. Any goods sold as intermediate goods would not get counted as part of the costs the disaster since they are already included in the final-goods computation. Individual economic losses, like wages, dividends, royalties, etc., would not be counted because they are already included in the expenditures. It is not immediately clear using the expenditures approach how difference in value due to the replacement of damaged equipment would be aggregated.

---

[17] Or other values, like book value, that have little relationship to either.

[18] Unless the company goes bankrupt, in which case there is some ambiguity.

The *income* approach would add up all lost profits, wages, royalties, and rents due to business interruption, and similarly the change in profits, wages, royalties, and rents due to the replacement of the damaged equipment. Lost dividends or capital gains to individuals or firms would not be counted as part of the loss from the disaster because those losses are already counted a part of lost profits.

The *value-added* approach would add up all the value added due to business interruption, and similarly the change in value added due to the replacement of the damaged equipment. Lost wages would likely be included, since labor adds value to the produced goods.

2.6.5 Resilience

There are very few data or measurement results directly on resilience. In general, we assume that resilient communities will minimize population loss, per capita economic loss, job loss, the amount of time people are without housing or unable to return to their homes, loss of essential services, etc. Furthermore we expect resilient communities to be able to establish a new 'normal' relatively quickly, and we expect that the more resilient the community the better the new 'normal' will be.

MCEER has developed a preliminary aggregate resilience index (MCEER 2010). They start with a series of quality-of-service (QOS) variables:

Which are normalized to fall in the range between 0 and 1, and where there is a separate QOS variable for each lifeline or variable of interest. Each individual QOS index can be treated as a resilience index in its own right, with some limitations. MCEER then combines the individual indices into a composite resilience index, using an independence assumption. For two indices, the composite index is.[19]

$$\frac{\cdot}{\cdot}$$

The approach is similar to composite indices used in other work. For example Meacham and Johann (2006) discuss a similar multiplicative index for rating risk in structures.

Given its intended use, any resilience index is effectively a welfare function, with all the strengths and weaknesses of welfare functions in general. As such it will be used for such purposes as recommending which lifelines should be repaired first, or for whom (e.g., Rose et al. 1997).

---

[19] An exposition of the general case is in the appendix.

There have been a handful of efforts directed at measuring elements of resilience. As already discussed, Kates and Pijawka (1977) and Vigdor (2008) discuss degree of recovery. Chang (2010) develops a methodology for measuring time to recovery, and applies it to Kobe, Japan after the earthquake. Rose et al. (2007) estimate the impact of business adjustment to minimize losses due to lifeline outages in a disaster. They include such measures as reliance on inventories, maintaining electrical generators (for resilience against electrical outages) and rescheduling of lost production. In their evaluation of the impact of a 2-week electrical outage in Los Angeles County, these forms of business resilience reduced losses by almost 80 %.

3 Summary and Recommendations for Further Research

3.1 Summary

Although there is a great deal of high-quality information available on resilience-related topics—hazard assessment, vulnerability assessment, risk assessment, risk management, and loss estimation—as well as disaster resilience itself, there is no central source of data and tools to which the owners and managers of constructed facilities, community planners, policy makers, and other decision makers can turn for help in defining and measuring the resilience of their structures and communities. The purpose of this document is to provide a survey of the literature and an annotated bibliography of printed and electronic resources that serves as that central source of data and tools to help readers develop methodologies for defining and measuring the disaster resilience of their structures and communities.

The survey of the literature on disaster resilience begins with a discussion of the key concepts—hazards, vulnerability, losses, and disasters—that feed into any methodology for defining and measuring disaster resilience. The discussion then focuses on the various definitions of disaster resilience that are found in the literature. For the purpose of setting the scope of the literature review found in this report, disaster resilience is defined as the ability to minimize the costs of a disaster, return to the *status quo*, and to do so in the shortest feasible time. A conceptual model of the "disaster cycle" is then introduced. The disaster cycle model consists of four stages: (1) response; (2) recovery; (3) preparedness; and (4) mitigation. The first two stages are "reacting to" a disaster, whereas the second two are "preparing for" a disaster. The conceptual model provides an ideal framework for highlighting the interactions between the four stages and how decision-making at the individual asset and community levels differ. Data and measurement issues associated with loss estimation are then described. These include critical analyses of case studies, insurance claims, direct measurement of losses, and survey-based methods. The discussion covers both direct losses due to damages to buildings and other constructed facilities and indirect losses due to business interruption and other second-order, disaster-related effects. The survey of the literature concludes with a mathematical treatment of two resilience-related topics—the optimal recovery path following a disaster and the theoretical basis for a multiplicative resilience index.

The annotated bibliography on disaster resilience covers technical reports, journal articles, software tools, databases, and web portals; it includes a synopsis of the key references as an overview and introduction to the materials abstracted. References are listed in alphabetical order by author, where the author may be a person, a company, an organization, or a government entity. An abstract is provided for each reference that summarizes the salient points of the reference. A URL is provided whenever a reference is available in electronic format.

3.2 Recommendations for Further Research

The background work for this report uncovered additional areas of research that might be of value to owners and managers of constructed facilities, community planners, policy makers, and other decision makers concerned with defining and measuring disaster resilience. Four recommendations for further research are put forward with an aim of strengthening the linkage between defining and measuring disaster resilience at the level of individual structures and at the community level. These four recommendations are concerned with: (1) a critical analysis of key structure types and their vulnerabilities, how these structure types interact with other physical infrastructure components in the event of a disaster, and how their performance would be improved if performance-based codes were implemented to address selected vulnerabilities; (2) an estimate of the incremental cost of implementing the proposed performance-based codes in the various structure types; (3) an estimate of the difference in expected property losses associated with implementing the proposed performance-based codes in the various structure types; and (4) an estimate of the difference in expected business interruption costs associated with implementing the proposed performance-based codes in the various structure types. The first three areas of research establish a foundation for measuring the disaster resilience of structures. The fourth area is an important link between the disaster resilience of structures and the disaster resilience of communities. The benefit of the proposed four-part research plan is that it provides the technical basis for using benefit-cost analysis of design and rehabilitation approaches to enhance the resilience of buildings.

3.2.1 Analysis of Key Structure Types

Buildings are systems of systems with many interactions and interdependencies among these systems. The performance of the structural system can have a direct impact on the performance of other systems (e.g., architectural, mechanical, electrical, etc.), as well as the utility infrastructure that can be housed inside, atop, or below the building. Thus the resilience of the structure has a substantial influence on the resilience of the building as a whole and in turn on the community. Furthermore, measures of resilience have been limited to assessing the extent of the adoption of current practices, standards, and codes for design and construction of buildings and infrastructure. This research has the goal of providing the technical basis for performance-based approaches to the design of new buildings or the rehabilitation of existing buildings to respond to natural hazards—earthquakes, community-scale fires, hurricane-strength winds and hurricane-borne storm surge, and tsunamis—and man-made hazards.

This research will involve the identification of a class of building types based on national and regional statistics. A set of hazards those building types face will then be formulated and a risk assessment conducted. The risk assessment will address different hazards using current building codes in place for the region of interest to identify vulnerabilities to those hazards. Researchers will then formulate a performance-based approach to modify the existing building code to address selected vulnerabilities.

3.2.2 Cost Analysis

Using national and regional statistics, data on building size, height, structural system, and primary use will be compiled for each building type selected for analysis in the first stage. These, "building characteristics" data, will be used to develop prototypical designs for the selected building types. Each design will have at least two alternative configurations: (1) reflecting current code provisions and (2) reflecting the proposed performance-based provisions to address selected vulnerabilities identified in the risk assessment. If appropriate, additional alternatives will be formulated that represent intermediate steps between the current code provisions and those associated with the performance-based approach. The prototypical building designs will be of sufficient detail that cost estimates can be developed that highlight the differences in the current code provisions and the proposed performance-based code provisions. Quantity take-offs and assemblies' data will be used to further analyze differences among alternative configurations.

Using national and regional statistics, weighting factors for each building type will be developed. These weighting factors reflect the proportion of a selected building type in a particular region of the country. National-level weighting factors might also prove useful. The purpose of the weighting factors is to develop regional-level estimates of the costs of implementing the improvements associated with the proposed performance-based provisions. Initial emphasis is placed on regional-level estimates because certain hazards affect some regions more than others. Consequently, the performance-based provisions recommended for one hazard/region combination may differ from those proposed elsewhere.

3.2.3 Analysis of Expected Property Losses

A variety of resources are available for estimating the expected property losses following a disaster. These include the HAZUS-MH software tool, the HAZUS-MH database, and other complementary databases. For purposes of this proposed research effort, HAZUS-MH is best suited. Because expected property losses may vary considerably across the different building types for a given hazard, a case study approach including both a baseline analysis (i.e., using most likely estimates of model input values) and a sensitivity analysis (i.e., reflecting uncertainty in the values of model inputs) is in order. The case study will select a geographical region, use the HAZUS-MH software tool to perform a series of hazard-specific simulations by varying the magnitude of the hazard, and analyze the property losses associated with different levels of hazard mitigation as reflected in selected building codes and their corresponding fragility curves. The specifications for the HAZUS-MH fragility curves will be examined to analyze ways to change key parameters and correlate those changes with the more stringent code requirements associated with the performance-based approach.

3.2.4 Analysis of Expected Business Interruption Costs

This study has demonstrated that business interruption costs are difficult to quantify. This is unfortunate since business interruption costs often equal or exceed property losses experienced following a disaster. Consequently, additional research is needed to analyze current and proposed approaches for measuring disaster-related, business-interruption costs. Better approaches for measuring business interruption costs—both at the microeconomic and macroeconomic levels—are crucial to a diverse set of stakeholders (e.g., insurance industry organizations, state and local emergency management agencies, and federal entities such as the Department of Homeland Security and the Bureau of Economic Analysis). Once a better methodology for measuring business interruption costs is available, it will be possible to use benefit-cost analysis of design and rehabilitation approaches to enhance the resilience of buildings.

Reducing business interruption costs creates an economic incentive for building owners if they occupy the structure. Conversely, if the building owner undertakes mitigation measures that will reduce expected business interruption costs of their tenants; they may be able to command a rental premium. Furthermore, business interruption costs are an important component of indirect, second-order, disaster-related costs. Thus, they are the first link in establishing the relationship between the disaster resilience of structures and the disaster resilience of communities.

Appendix A   Disaster Resilience: An Annotated Bibliography

The literature on disasters and disaster resilience is vast and draws from a wide number of fields including climatology, geography, geology, sociology, political science and economics. Its objectives range from answering purely scientific questions to determining optimal responses to overt political advocacy. The bibliography on the subsequent pages is only a sampling of it.

Tierney et al. (2001) probably provide the best overview of the field. They have chapters on preparedness, individual and group behavior in disasters, organizational and government response in disasters, and factors influencing preparedness and response.

Waugh and Tierney (2007) write a textbook on emergency management and response to disasters. Its intended audience is local government officials who are tasked with developing and implementing emergency response plans and managing the local response to disasters. This book is different from the one by Tierney et al. in that this one is a more practical and down-to-earth guide to disaster management and response while the previous one provides a more general and theoretical view of disaster response.

J. Haas, R. Kates, M. Bowden, eds. (1977) are the seminal work in understanding disaster response and recovery. Nearly all subsequent work in understanding response and recovery reference them. Much of the subsequent work on recovery uses the framework established by Haas et al. However, many of the researchers following them disagree with the Haas framework (e.g., Berke et al. 2008, Bolin and Bolton 1986). The problem is two-fold. (1) The Haas framework is more of a way of organizing and interpreting information than it is really an empirical result. (2) Subsequent researchers are interested in different questions. They are interested in differences in impact and recovery far more than they are interested in the degree of recovery.

Miles and Chang (2003) and Chang (2010) develop a set of simulations and models of the recovery process. The work is preliminary, but it provides a foundation on which further work can be built.

Bruneau et al. (2003) develop a framework to quantitatively assess disaster resilience. They have continued and refined that work since at MCEER.

There have been a series of studies aimed at estimating indirect costs of disasters, including in particular Rose and Liao (2005) who develop Computable General Equilibrium methods for estimating disaster losses and apply them to the hypothetical case of an earthquake in Portland Oregon. Also West and Lenze (1994) who carefully develop the data that would be needed to model disaster impacts in regard to Hurricane Andrew.

Rose et al. (2007) evaluate the benefits versus the costs of FEMA Hazard Mitigation Grants, and find that the benefits exceed the costs of the grants by an average magnitude of about 4:1.

B. Adams, C. Huyck, B. Mansouri, R. Eguchi, M. Shinozuka, 2004. "Application of High-Resolution Optical Satellite Imagery for Post-Earthquake Damage Assessment: The 2003 Boumerdes (Algeria) and Bam (Iran) Earthquakes" MCEER Research Progress and Accomplishments 2003-2004: 173-186.

> This is a description of a near-real-time damage assessment algorithm based on the comparative analysis of remote sensing images acquired before and after the event. They also develop methods for integrating these techniques into field-based reconnaissance activities.

A. Amendola, Y. Ermoliev, T. Ermoliev, V. Gitis, G. Koff, J. Linerooth-Bayer, 2000. "A Systems Approach to Modeling Catastrophic Risk and Insurability" Natural Hazards 21 (2-3): 381-393.

> They develop a methodology for estimating risk of damage, and in particular risk of insolvency of an insurance company (or companies). They then apply it to a hypothetical earthquake in the Irkutsk region of Russia. Their approach is analytically intractable, so they use a Monte Carlo simulation and accounting for the interdependencies in the risks.

American Society of Civil Engineers, 2009. *2009 report card for America's infrastructure.* Reston Va.: American Society of Civil Engineers.

> They evaluate the levels of infrastructure investment in the U.S., including Aviation, Bridges, Dams, Drinking Water, Energy, Hazardous Waste, Inland Waterways, Levees, Public Parks and Recreation, Rail, Roads, Schools, Solid Waste, Transit, and Wastewater. All are judged to be underfunded and their resilience is considered to be low.

P. Berke, J. Kartez, D. Wenger, 2008. "Recovery after Disaster: Achieving Sustainable Development, Mitigation and Equity" *Disasters* 17 (2): 93-109.

> This paper reviews key findings of the disaster recovery literature, but forms a perspective different from much of that literature. The authors are interested particularly in equity, mitigation, and sustainable development. They are especially interested in the impact of redevelopment planning and institutional cooperation on these outcomes.
>
> They review Haas' model of recovery (infra) and argue against it. Their main objection is that his model is not useful for what they are interested in. As mentioned, they are interested in issues of equity and sustainability, and Haas' model of recovery does not address those topics.
>
> They propose to replace Haas' model with what is in effect a model of disaster aid rather than a model of recovery. Their model is centered around (re)development planning. They argue that good aid has four characteristics:
>
> 1. Goodness of fit to local needs and opportunities;

2. Reliance on local resources.

3. Local knowledge about how to obtain disaster assistance.

4. Flexibility and willingness to learn in process.

Bin, Okmyung, Tom Crawford, Jamie Kruse, and Craig Landry. 2006. "Valuing Spatially Integrated Amenities and Risks in Coastal Housing Markets." Working Paper.

> Separating the value of coastal amenities from the negative value of risk from coastal storms is extremely difficult using hedonic methods due to the high correlation between the two. In this study, they construct a three-dimensional measure of view, accounting for natural topography and built obstructions, that varies independent of the risk to disentangle these spatially integrated housing characteristics. A spatial hedonic model is developed to provide consistent estimates of the willingness to pay for coastal amenities and risk. Their results suggest that such techniques can be successful in isolating risk from amenities on the coast.

R. Bolin, P. Bolton, 1986. *Race, Religion, and Ethnicity in Disaster Recovery*. Institute of Behavioral Science, University of Colorado.

> They study four specific disasters: the Coalinga Earthquake (1983); Paris, TX Tornado (1982); Wasatch Front Floods, UT (1983); and Hurricane Iwa, Kauai HI, (1982). They are primarily interested in the differences in recovery between ethnic groups. They study the psychological resiliency of the various groups and factors affecting it; (differential) degree of impact, Economic and Social Recovery Time, and (differential) utilization of aid.

R. Bolin, L. Stanford, 1991. "Shelter, Housing and Recovery: A Comparison of U.S. Disasters" *Disasters* 15 (1): 24-34.

> They examine the issues associated with the temporary sheltering and housing of victims after natural disasters in the United States. They are particularly interested in differential access to shelter and housing aid according to social class, ethnicity and related demographic factors; the relationship between post-disaster shelter and housing and long-term recovery; the role of social support networks in the sheltering of victims; and the implications of the research for the provision of shelter and housing aid after disasters.

> Two observations from the paper should be mentioned. First, in effect an earthquake erases the bottom tier of housing, either by directly destroying it or by forcing it to be upgraded and therefore priced out of the bottom tier housing market. Second, the more aid is devoted to temporary housing, the more time it tends to take for permanent housing to be completed.

K. Borden, S. Cutter, 2008. "Spatial patterns of natural hazards mortality in the United States" *International Journal of Health Geographics* 7 (1): 64.

They use data from the SHELDUS database (augmented with some supplemental data) to estimate the spatial patterns of natural hazard mortality at the county-level for the U.S. from 1970–2004 using a combination of geographical and epidemiological methods. Note that this excludes Katrina.

They find that chronic everyday hazards such as severe weather (summer and winter) and heat account for the majority of natural hazard fatalities. The regions most prone to deaths from natural hazards are the South and intermountain west, but sub-regional county-level mortality patterns show more variability.

The greatest mortality is due to heat, with generic bad weather (excluding hurricanes, tornados, floods, and lightning) making up the majority of the deaths. Hurricanes and earthquakes are minor players in deaths.

Event Deaths		Pct
Heat/Drought 3,906		19.57 %
Severe Weather	3,762	18.85 %
Winter Weather	3,612	18.10 %
Flooding 2,788		13.97 %
Tornado 2,314		11.59 %
Lightning 2,261		11.33 %
Coastal 456		2.28 %
Hurricane/Trop. Storm	304	1.52 %
Geophysical 302		1.51 %
Mass Movement	170	0.85 %
Wildfire 84		0.42 %

S. Brody, S. Zahran, W. Highfield, H. Grover, A. Vedlitz, 2008. "Identifying the impact of the built environment on flood damage in Texas" *Disasters* 32 (1): 1–18.

This study examines the relationship between the built environment and flood impacts in Texas. They regress (log) property damage resulting from 423 flood events between 1997 and 2001 at the county level. The regressors include several natural environment variables, the number of dams in the county, wetland alteration in the county, percent impervious surface in the county, and median household income.

Their results suggest that naturally occurring wetlands play a particularly important role in mitigating flood damage. They also reproduce the standard result that impervious surface increases the amount of flood damage.

Bruneau, Michel et al. 2003. "A Framework to Quantitatively Assess and Enhance the Seismic Resilience of Communities." *Earthquake Spectra* 19:733-752.

> This paper presents a conceptual framework to define seismic resilience of communities and quantitative measures of resilience that can be useful for a coordinated research effort on enhancing this resilience. It is an early version of work that is still underway at MCEER.
>
> In this framework there are 4 dimensions of resilience: Technical (survivability of physical systems); Organizational (Capacity of key organizations to respond); Social; and Economic. They are divided into two groups, Technical and Organizational, which focus primarily on the protective side, and Social and Economic which focus primarily on the protected side.
>
> They also consider 4 'aspects' that they intersect with their 4 dimensions: Robustness; Redundancy; Resourcefulness; and Rapidity.

R. Burby, R. Deyle, D. Godschalk, R. Olshansky, 2000. "Creating Hazard Resilient Communities through Land-Use Planning" *Natural Hazards Review* 1 (2): 99-106.

> They argue that land-use planning is the single most promising approach for bringing about sustainable hazard mitigation. This article describes the essential elements of land-use planning for hazard mitigation. It highlights important choices involved in formulating planning processes, undertaking hazard assessments, and crafting programs to manage urban development so that it is more resilient to natural hazards.
>
> They suggest that the following planning tools are usable for disaster mitigation:
>
> - Building standards
>
> - Development regulations: i.e., zoning.
>
> - Critical- and public-facilities policies: e.g., don't build the hospital on the fault.
>
> - Land and property acquisition: e.g., turn the flood plain into a park.
>
> - Taxation and fiscal policies: e.g., tax the properties on the wildland interface.
>
> - Information dissemination.
>
> They also offer the following Principles for development management:
>
> - Use clear and authoritative maps of the hazard, with an emphasis on clear.
>
> - Link clear and realistic design guidelines to the maps: so everyone knows what it means for them.

- Ensure that hazard-free land is available for development:

- If trying to rearrange or restrict land uses in hazardous areas, do so before the land is subdivided.

- Offer incentives to encourage developers to locate projects outside of hazardous areas and to adopt hazard mitigation measures that exceed those required by law.

- If hazardous land is subdivided and built out, be prepared to purchase selected properties.

- Use project-specific design approaches.

- Use the post-disaster window of opportunity to encourage individual owners to retrofit or relocate.

S. Chang, C. Chamberlin, 2004. "Assessing the Role of Lifeline Systems in Community Disaster Resilience" *MCEER Research Progress and Accomplishments 2003-2004*: 87-94.

> This is a preliminary effort to model the impacts of infrastructure damage on business. The key innovation here is to take into account the overlap in damage. They estimate the damage to infrastructure systems (including—or perhaps especially—overlaps). Then they distribute the 'disruptiveness' randomly across affected businesses. Finally, they estimate probabilities of business closure based on how many 'disruptive' interruptions there are. Finally they propose to apply the model to the Los Angeles Department of Water and Power (LADWP).

S. Chang, 1983. "Disasters and Fiscal Policy: Hurricane Impact on Municipal Revenue" *Urban Affairs Review* 18 (4): 511-523.

> This study investigates the fiscal impact of Hurricane Frederic on the city of Mobile, Alabama. The hurricane caused damages of about $1.6 billion in property and other losses in the state, led to an influx of $670 million in recovery funds, and resulted in a $2.5 million increase in municipal revenue for the next 12 months. The long-term impact was negative, however, because secondary effects overwhelmed the revenue effect. For example, flooding as a result of tree destruction from the hurricane cost more than revenue gains from disaster aid.

S. Chang, 2010. "Urban disaster recovery: a measurement framework and its application to the 1995 Kobe earthquake" *Disasters* 34 (2): 303-327.

> This continues Chang's recent work on disaster recovery by applying it to Kobe. The paper provides a framework for assessing empirical patterns of urban disaster recovery through the use of statistical indicators.

She discusses several different possible meanings for 'recovery,' and provides measures for each of them in the Kobe analysis. There is an effort also to filter out exogenous influences unrelated to the disaster, and to make comparisons across disparate areas or events.

It is applied to document how Kobe City, Japan, recovered from the catastrophic 1995 earthquake. Findings reflect the results of Haas (1977) in a number of ways. For example, while aggregate population regained pre-disaster levels in ten years, population had shifted away from the older urban core and become more suburban. Longer-term trends in business mix were generally accelerated. In particular there were sectoral shifts toward services and large businesses. There was a three to four year temporary boost in economic activity from reconstruction activities,

The city did not seem to fully recover. While population returned to pre-disaster levels within ten years (from a 6 % decline), economic activity eventually settled at a level some ten per cent below pre-disaster levels. Similarly there were substantial losses of port activity that do not seem to be presaged by longer term trends.

She grades recovery based on "time to recovery" with two standards: return to previous level, and "attainment of a new 'normal.'"

Return to previous level is intuitive and straight-forward. Return to new normal is defined here. She picks a pre-disaster comparison period (in her case, four years), and selects the greatest variation in that period. So:

$$max$$

Then Normal is judged to be reached when annual variation becomes less than that critical amount.

S. Chang, A. Falit-Baiamonte, 2002. "Disaster vulnerability of businesses in the 2001 Nisqually earthquake" *Global Environmental Change Part B: Environmental Hazards* 4 (2-3): 59-71.

This paper examines the impacts of the February, 2001, Nisqually earthquake on businesses. The study investigates the extent of losses, patterns of disparities, and underlying loss factors.

Results showed that business losses were much greater than what standard statistical data would imply. They find that businesses are largely self-insured: 80 % cover losses out of their own pocket. They note that since most of the losses were self-financed, they tend to be uncounted in most official statistics. That suggests business losses are almost entirely uncounted.

In this case, SMALL and RETAIL were strongly correlated with loss. Preparedness does not have any significant relationship at all to loss. Preparedness and Vulnerability correlate highly negatively. That is, Big Non-Retail Building owners prepare, while small retail renters do not.

S. Chang, M. Shinozuka, J. Moore, 2000. "Probabilistic Earthquake Scenarios: Extending Risk Analysis Methodologies to Spatially Distributed Systems" *Earthquake Spectra* 16 (3): 557-572.

> This paper proposes a methodology by which probabilistic risk analysis methods can be extended to the assessment of urban lifeline systems. The problem with its previous application to lifeline systems is that systems such as highway networks, electric power grids, and regional health care delivery systems, are geographically distributed, and the failure at one point degrades the ability of the entire system to function. So the spatial correlation between earthquake ground motion across many sites is important in determining system functionality.
>
> The methodology developed in this paper first identifies a limited set of deterministic earthquake scenarios and evaluates infrastructure system-wide performance in each. It then assigns hazard-consistent probabilities to the scenarios in order to approximate the regional seismicity. The resulting probabilistic scenarios indicate the likelihood of exceeding various levels of system performance degradation.
>
> A demonstration for the Los Angeles study area highway network suggests that there is roughly a 50 % probability of exceedance of Northridge-level disruption in 50 years. This methodology provides a means for selecting representative earthquake scenarios for response or mitigation planning.

Chapman, Robert, and Amy Rushing. 2008. *Users Manual for Version 4.0 of the Cost-Effectiveness Tool for Capital Asset Protection*. NIST IR 7524. Gaithersburg, MD: NIST.

> This paper describes the CET 4.0 software tool. It is intended to implement the ASTM 2506 protocol for evaluating the cost effectiveness of proposed disaster mitigation measures. It takes as inputs costs of the various proposed alternatives, including O&M costs; costs associated with relevant disaster events, and the probabilities associated with the events. It is capable of handling uncertainty in costs of likelihood of adverse events using Monte Carlo methods.

S. Cho, P. Gordon, J. Moore, H. Richardson, M. Shinozuka, S. Chang, 2001. "Integrating Transportation Network and Regional Economic Models to Estimate the Costs of a Large Urban Earthquake." *Journal of Regional Science* 41 (1): 39-65.

> In this paper they summarize an integrated model of losses due to earthquake impacts on transportation and industrial capacity, and how these losses affect the metropolitan economy. They divide damage into four groups: Structural Losses, Direct Business Losses, Indirect Business Losses, and Costs of Increased Transportation Time. Indirect

business losses are due entirely to transportation problems; e.g., supplies and labor being unable to get to work sites, etc.

Preliminary results are summarized for a magnitude 7.1 earthquake on the Elysian Park blind thrust fault in Los Angeles. Here, costs of increased transportation time are on an order with structure losses. Indirect business losses are on an order with direct business losses. The amount of indirect losses are underestimated since there are clearly other sources of indirect losses than transportation.

H. Cochrane, 2004. "Indirect Losses from Natural Disasters: Measurement and Myth." pp. 37-52. In Okuyama, Yasuhide, and Stephanie Chang, eds. *Modeling Spatial and Economic Impacts of Disasters*. Springer.

He runs a series of HAZUS simulations to estimate the amount of indirect losses from various natural disasters. Based on those simulations he observes:

- There is no indirect loss if all sectors of the economy are equally damaged.

- Indirect losses are the predominant loss if a small critical sector goes down.

- Indirect losses are negative if there is ample excess capacity, and reconstruction is financed by outside sources, but for the nation they must be positive.

- Indirect losses are temporarily negative if reconstruction is financed out of savings or borrowing.

- Indirect losses are less sensitive to economic structure than to damage pattern, degree of integration, preexisting conditions, and financing source.

Collins, D., and S. Lowe. 2001. "A Macro Validation Dataset for U.S. Hurricane Models." in *Casualty Actuarial Society Forum*. Arlington, VA: Casualty Actuarial Society.

They have developed a macro validation data set for the U.S. hurricane history between 1900 and 1999. This should enable a lay person to compare the overall results of modern hurricane models to an historical record.

The macro validation dataset consists of the aggregate insured losses from hurricanes affecting the continental United States from 1900 through 1999. The historical losses in each county have been "trended"—adjusted from the conditions at the time to those existing today. The trending reflects not only estimated changes in price levels, but also estimated changes in the value of the stock of properties and contents, and changes in the insurance system.

L. Comfort, 1999. *Shared Risk: Complex Systems in Seismic Response*. Pergamon.

> She is primarily interested in adaptive response to earthquakes. Communication, flexibility and adaptability are her big issues. In this she continues the work of Quarantelli, Wenger, and the others at the Disaster Research Center who have been identifying communications as a major problem with disaster response for many decades.
>
> Here case studies ranged from cases where communities "failed to address the larger goal of seismic risk reduction," and did fairly poor jobs at relief and restoration. On the other end there were cases where response was highly adaptive. There were frequently problems with the response, and the more severe the disaster the greater the problems.
>
> One key observation is that local governments and organizations are barely in evidence in any of the 'non-adaptive' case studies. This contrasts with her 'operative-' and 'auto-adaptive systems,' where the response was primarily local.
>
> Characteristics of good Response based on her comments:
>
> 1. Strong, Independent, Capable local responders. In the case of minor 'disasters,' they will be able to take care of the problem themselves. In large 'no warning' disasters, their resources may not be sufficient to the task, but they will still be the only ones on site for hours or days, so their reaction is still critical. Note that this includes having the locals highly trained and skilled.
>
> 2. Ability to scale up. In large disasters, the ability to call in additional resources promptly becomes critical.
>
> 3. Ability to coordinate response. The ability to put resources where they are needed.
>
> Two observations about the case studies. First, responses are best where they are local. That is, the cases where the locals did the main work came out best. Second, communication is important when the scale of the disaster begins to overwhelm local resources. In Whittier Narrows and Loma Prieta, local resources were (mostly) sufficient to the task so the communications problems she identifies are not important to the outcome. In Hanshin and Erzincan, where the scale of the disaster overwhelmed local resources, lack of communication between local authorities and national/international aid sources significantly slowed the response.

L. Comfort, T. Haase, 2006. "Communication, Coherence, and Collective Action: The Impact of Hurricane Katrina on Communications Infrastructure" *Public Works Management & Policy* 10 (4): 328-343.

> Here they use the methodology from Comfort (1999) for Katrina, and perform some additional analysis.

Coughlin, K., E. Bellone, T. Laepple, S. Jewson, and K. Nzerem. 2009. "A Relationship between all Atlantic Hurricanes and those that make landfall in the U.S." *Quarterly Journal of the Royal Meteorological Society* 135:371-379.

They first investigate whether there are statistical breaks in the hurricane record. They find that there are three breaks: 1945, which probably representing improved data collection; 1967, and 1994. The period from 1967 – 1994 is a well-known lull in Atlantic hurricane activity.

Their results suggest that the proportion of Atlantic hurricanes that hit the U.S. is a random process with fixed distribution. Specifically, about 1/3 of all Atlantic hurricanes hit the U.S., and the proportion seems to be constant over time.

S. Cutter, B. Boruff, L. Shirley, 2003. "Social Vulnerability to Environmental Hazards" *Social Science Quarterly* 84 (2): 242-261.

Their objective is to develop an index of social vulnerability to environmental hazards based on readily available county-level socioeconomic and demographic data.

They pick about 40 variables that people believe are associated with vulnerability to risk. They then do a principle-components analysis on the variables, and come up with about 10 that explain most of the variation in their associated variables. To determine the Social Vulnerability Index (SoVI), they do an unweighted sum of the variables. Then they break that into 5 bins based on standard deviation of the sum.

They find that there are some distinct spatial patterns in the SoVI, with the most vulnerable counties clustered in metropolitan counties in the east, south Texas, and the Mississippi Delta region.

They do not attempt to correlate any of the factors with actual disaster impacts. Nor is county level wealth corrected for regional price variations. The index is *negatively* correlated with presidential disaster declarations. However, disaster declarations have a strong political component.

Dahlhamer, James, and Kathleen Tierney. 1998. "Rebounding from Disruptive Events: Business Recovery Following the Northridge Earthquake." *Sociological Spectrum* 18:121-141.

They develop a model of business recovery by drawing from existing research on disaster recovery and on organizational survival in non disaster contexts and test it by using data collected from a stratified random sample of 1 110 Los Angeles area firms affected by the 1994 Northridge earthquake.

Key predictors of recovery were size (positively), degree of disruption (negatively), Shaking Intensity (negatively), and whether the business was a recipient of post-disaster aid (negatively).

The negative impact of aid is probably a result of self-selection effects (only the worst-off businesses obtained aid) and problems such as higher debt caused by the aid itself.

They argue that 'ecological factors' are important to business recovery. For example, a business that was undamaged and experienced no interruptions may still do poorly if it is located in a district where most of the surrounding businesses were destroyed.

Davidson, Rachel, and Kelly Lambert. 2001. "Comparing the Hurricane Disaster Risk of U.S. Coastal Counties." *Natural Hazards Review* 2:132-142.

This paper describes the Hurricane Disaster Risk Index (HDRI), a composite index developed to compare the risk of hurricane disaster in U.S. coastal counties. The HDRI is a composite of four separate sub-indices: Hazard, Exposure, Vulnerability, and Emergency Response and Recovery Capability. The factors contributing to the sub-indices are weighted according to expert judgment. The sub-indices are combined multiplicatively to get the HDRI. The paper includes an analysis of 15 sample counties.

A. Dlugolecki, 2008. "An overview of the impact of climate change on the insurance industry." pp. 248-278. In Diaz, Henry, and Richard Murnane, eds. *Climate extremes and society*. Cambridge University Press.

This chapter focuses on the types of extreme weather and climate events that are important to property insurers, and it considers evidence on how those risks have been changing and how they might change in the future with climate change.

He does a superficial analysis of insured losses due to flood damage in the UK where insurance penetration for flooding is high. He finds a significant upward trend in insured flood losses. The analysis is very incomplete and has some flaws (for example, he does not correct for trends in Real GDP or insurance penetration), but it seems likely that there will be an upward trend even after correcting for those items.

He believes there is good ground to argue that climate change is already affecting the risks, although it is not the only factor that has caused change. He argues that weather losses will increase rapidly, and cites others' models to support this idea. In particular, he suggests that extreme losses could increase very rapidly.

A. Dodo, R. Davidson, N. Xu, L. Nozick, 2007. "Application of regional earthquake mitigation optimization" *Computers & Operations Research* 34 (8): 2478-2494.

and

N. Xu, R. Davidson, L. Nozick, A. Dodo, 2007. "The risk-return tradeoff in optimizing regional earthquake mitigation investment" *Structure and Infrastructure Engineering* 3 (2): 133–146.

These papers introduce a stochastic optimization model developed to help decision-makers understand the risk-return tradeoff in regional earthquake risk mitigation, and to help state and local governments comply with the Disaster Mitigation Act of 2000 requirement that they develop a mitigation plan. A case study for Central and Eastern Los Angeles illustrates an application of the model. Results include a graph of the tradeoff between risk and return, quantification of the relative contributions of each possible earthquake scenario, and discussion of the effect of risk aversion on the selection of mitigation alternatives.

They develop a method to compute the optimal set of seismic upgrades to buildings in a large area. The mitigation alternatives considered are structural upgrading policies for groups of buildings. They use the HAZUS model to simulate damage, and they design a linear program to estimate how much to spend on pre-earthquake mitigation, and which of the many possible mitigation activities to fund so as to minimize overall risk.

The model is intended to be used as a tool to support the public regional mitigation planning process. In any realistic application, the model would include millions of variables, thus requiring a special solution method. This paper focuses on two efficient solution algorithms to solve the model.

K. Donaghy, N. Balta-Ozkan, G. Hewings, 2007. "Modeling Unexpected Events in Temporally Disaggregated Econometric Input-Output Models of Regional Economies" *Economic Systems Research* 19 (2): 125-145.

Their objective is to develop the framework for Regional Econometric Input-Output Models (REIMs) at high spatial resolution. Their main focus is on developing a structural econometric model for estimating the Input-Output table at the core of the model. They conclude by estimating a hypothetical example where a single industry is shut down for four weeks, and estimate the impacts on the economy at 2-week intervals.

Downton, Mary, J Zoe Barnard Miller, and Roger Pielke. 2005. "Reanalysis of U.S. National Weather Service Flood Loss Database." *Natural Hazards Review* 6:13-22.

and

M. Downton, R. Pielke, 2005. "How Accurate are Disaster Loss Data? The Case of U.S. Flood Damage" *Natural Hazards* 35 (2): 211-228.

These papers are an analysis of the accuracy of the flood loss data collected by the U.S. National Weather Service since about 1933. The reanalyzed data are posted at http://www.flooddamagedata.org/.

These papers use historical flood damage data in the U.S. to evaluate disaster loss data. The NWS estimates are obtained from diverse sources, compiled soon after each flood event, and not verified by comparison with actual expenditures. These papers presents results of a comprehensive reanalysis of the scope, accuracy, and consistency of NWS damage estimates from 1926 to 2000.

1. Damage estimates for individual flood events are often quite inaccurate. Although, the larger the event, the more accurate (in log terms) the estimate.

2. With the caveats mentioned below, the data seem to be unbiased.

3. The data are bottom-censored, that is, any event smaller than a specific size is not included in the database. That means the totals are probably biased down.

4. Most of the damage occurs in a few very large events. So, for example, in Pennsylvania (which has the highest total damage in the database), something like 75 % of the total damage (over 75 years) occurred in a single flood event.

R. Ellson, J. Milliman, R. Roberts, 1984. "Measuring the Regional Economic Effects of Earthquakes and Earthquake Predictions" *Journal of Regional Science* 24 (4): 559-579.

They build a predictive model of the Charleston, SC area economy. They simulate the area with and without a serious earthquake. Their model explicitly takes into account the types of supply-side constraints one would expect to encounter in a natural disaster.

Estimates of deaths and losses of capital are more or less ad hoc. The simulations assume a global death and loss rate based on analogy with the Los Angeles area. They assume that there is no external aid.

They run four simulations: a baseline (no earthquake) simulation, an unanticipated earthquake, a predicted earthquake, and a false-alarm prediction.

The simulations are ranked in terms of total losses in the expected order. However, there are significant distributional differences in the different simulations. For example, the unanticipated disaster results in large increases (relative to the other simulations) in labor and proprietor income, largely due to the resulting construction boom.

Y. Ermoliev, T. Ermolieva, G. MacDonald, V. Norkin, A. Amendola, 2000. "A system approach to management of catastrophic risks" *European Journal of Operations Research* 122: 452-460.

They are interested in understanding the behavior and optimal choice of an insurance company facing correlated (catastrophic) risk. In their models, firms are bankruptcy averse. Firms have the option of buying reinsurance or catastrophe bonds to distribute their risks.

The loss and optimization functions are discontinuous, which makes optimizing over them particularly difficult. They discuss Stochastic Dynamic Optimization techniques that will solve such problems.

FEMA. 2005. *Risk Assessment: A How-To Guide to Mitigate Potential Terrorist Attacks Against Buildings*. FEMA 452. Washington DC: FEMA.

This is a how-to guide for the development and evaluation of alternatives for mitigation against terrorist attacks. The guidebook uses five steps to complete the process. The first step is to list and assess possible threats. Second, they evaluate the value of the protected asset. Value in their case means evaluating how serious the consequences would be if the building were destroyed or taken out of service. They also evaluate the value of building and security components. Third they estimate vulnerability of the building. Fourth is the risk assessment, which combines the results of the first three into a single index value. Finally (step five), they develop a set of mitigation options, by asking what measures will reduce the risk rating, how effective the measures are and how much they cost (in both monetary and non-monetary terms).

A. Fothergill, 1998. "The Neglect of Gender in Disaster Work: An overview of the literature." pp. 11-26. In Enarson, Elaine, and Betty Morrow, eds. *The gendered terrain of disaster: through women's eyes*. Westport Conn.: Praeger.

She makes several observations about the behavior of women during the disaster cycle, and how they differ from men:

- Women are more risk averse. They are more likely to evacuate. But they are less likely to take mitigation measures.

- Women are more likely to buy safer homes (and insist that builders follow building codes).

- Women are more likely to hear warnings (due to better peer networks), and take them seriously. Men are less likely to take the warnings seriously. Married men often evacuate at the behest of their wives, even though they themselves would prefer not to.

- The results on mortality are mostly taken from the third world, since there are very few deaths in the U.S. to base results on. In general, if the disaster is outdoors (lightning, wind, etc) men die more. If the disaster is indoors (earthquakes, flooding, etc.) women die more.

H. Friesma, J. Caporaso, G. Goldstein, R. Linberry, R. McCleary, 1979. *Aftermath: Communities after natural disasters*. Beverly Hills, Calif: Sage.

They look at several communities after disasters and try to do an analysis of trends to see if there are any long-term effects of the disaster. The disasters are mostly small relative to the size of the community.

- They had a hard time finding an effect on unemployment due to the disasters.

- Effects on numbers of small businesses are estimated by eye, and the results are ambiguous. In some communities there is a decline, in others an increase. In almost all cases, the effect is small in comparison to the larger trend.

- There is no evidence of impact on total sales at the annual level.

- In two communities there is strong evidence of a temporary drop in divorce that is not made up later. That is, some families that would have divorced simply never did due to the disaster.

- The effects on crime are ambiguous.

- Population figures are only reliable for one community. There they suggest a brief decline (less than 2 years) followed by recovery to trend. There is some evidence of dispersal of the population to the fringes a la Haas.

E. Glaeser, J. Gottlieb, 2009. "The wealth of cities: agglomeration economies and spatial equilibrium in the United States" *Journal of Economic Literature* 47 (4): 983–1028.

This is a survey article and general introduction to urban economics.

Empirical research on cities starts with a spatial equilibrium condition: workers and firms are assumed to be indifferent across space. This condition implies that research on cities is different from research on countries, and that work on places within countries needs to consider population, income, and housing prices simultaneously.

- Not only do incomes vary across locations (e.g., San Francisco has a much higher per capita income than Brownsville), but so do prices. After factoring in prices, the spatial differences in income are less variable.

- Nevertheless, people express a mild preference for rural as opposed to urban living in that they are willing to accept a decrease in standard of living to live in the country.

- Mobility and spatial equilibrium suggest that poverty in cities is less a product of cities being bad (since people can leave). Rather it is a product of cities providing better services to the poor (and thus attracting them).

D. Godschalk, T. Beatley, P. Berke, D. Brower, E. Kaiser, 1999. *Natural Hazard Mitigation: recasting disaster policy and planning*. Washington DC: Island Press.

The basic questions they ask are whether state mitigation plans meet the requirements of the Stafford Act, whether they meet basic standards of quality, and whether they are consulted and implemented in disaster recovery. In their six case studies, mitigation plans were largely unconsulted in the recovery period and in deciding what mitigation (if any) to actually implement.

P. Gordon, J. Moore, H. Richardson, 2002. *Economic-Engineering Integrated Models for Earthquakes: Socioeconomic Impacts*. Pacific Earthquake Engineering Research Center: College of Engineering: University of California, Berkeley.

They are interested in how the impacts of an earthquake differ depending on (mainly) income. The Southern California Planning Model-2 (SCPM-2) is used to model the economic impacts of a hypothetical earthquake on Los Angeles's Elysian Park fault.

Census data on occupation are used to distribute these impacts across income groups within each city in the region. This permits the impacts of such an earthquake, and potential mitigation programs, to be assessed in terms of city-specific changes in income equity measures.

In the Elysian Park scenario, the data suggest that the well-off suffer a (slightly) disproportionate share of the damage. That is, this is one disaster that strikes the wealthy more than the poor. They suggest that information is the main reason. There is no readily available information to say where the risks are, so the wealthy can't really avoid them.

P. Gordon, H. Richardson, B. Davis, 1998. "Transport-related impacts of the Northridge earthquake" *Journal of Transportation and Statistics* 1 (2): 21–36.

This research estimates the transport-related business interruption impacts of the 1994 Northridge earthquake using a spatial allocation model, the Southern California Planning Model (SCPM), and surveys of businesses and individuals.

Total business interruption losses are estimated at more than $6.5 billion, sizeable but much smaller than total structural damage (over $25 billion), with an associated job loss of 69,000 person-years.

The four types of transport-related interruptions (commuting, inhibited customer access, and shipping and supply disruptions) totaled more than $1.5 billion, or 27.3 % of all local business interruptions, with a job loss of more than 15 700 person-years. In addition, there were commuting travel time losses of at least $33 million and some dislocation of shopping patterns and frequencies. These losses would have been much higher had it not been for the substantial redundancy in Los Angeles' highway system.

P. Gordon, H. Richardson, 1995. *The Business Interruption Effects of the Northridge Earthquake*. Lusk Center Research Institute: University of Southern California.

They do a preliminary analysis of aggregate economic statistics for the LA area surrounding the Northridge quake and find no discernable impact. There are issues with such an analysis, but what that does is place an upper limit on the amount of business interruption that occurs.

Then they use a survey of businesses in the impact zone with the idea of feeding it into the Southern California Planning Model to distribute regional impacts. Among other things, they find:

- Total business interruption losses are about $6 billion, which is in the range of 20 % of total losses. This is not small, but it is less than earlier hypothetical studies had suggested. About 2/3 of the losses are outside the direct zone of impact, and about ¼ of the business losses are outside the region.

- Employment losses (not job losses) amounted to approximately 75,500 person years or about 1.2 % of the total employment in the area.

- The support for the usual finding that small businesses are more vulnerable is weak.

They do not account for survivor bias, or for forward impacts.

P. Guimaraes, F. Hefner, D. Woodward, 1993. "Wealth and Income Effects of Natural Disasters: An Econometric Analysis of Hurricane Hugo" *Review of Regional Studies* 23: 97-114.

They use a regional economic model, calibrated to the period before Hurricane Hugo, as the baseline for comparison. The model takes into account the 1991 recession and the impacts of the Gulf War. Deviations from the model are assumed to be due to the hurricane.

They "found that South Carolina's total personal income suffered a major drop in the third quarter of 1989 due to the hurricane. ... To a large extent, however, the personal income effect is an accounting, not economic artifact. ... The decline in South Carolina's personal income in the third quarter of 1989 largely reflected damage to structures caused by the hurricane and figured in the calculation of imputed rental income."

"One unexpected result to emerge from the analysis is the length of time the hurricane affected the economy. ... [T]here is a negative economic after-shock two years after the disaster. Apparently, residents and businesses undertake construction projects that may have been part of normal maintenance, along with storm damage repair. This boosts construction income in the quarters following the disaster, but dampens it in later quarters."

H. John Heinz III Center for Science, Economics, and the Environment., 2000. *The hidden costs of coastal hazards: implications for risk assessment and mitigation*. Washington D.C.: Island Press.

They define a "Disaster Resilient Community" as "a community built to reduce losses to humans, the environment, and property as well as the social and economic disruptions caused by natural disasters. It is the safest community that can be designed and built given the status quo and budget and resource constraints."

They divide losses into four categories:

- Built environment

- Business environment: mostly 'indirect' business interruption losses. But they would also include things like loss of inventory, tools, equipment, etc., in this category

- Social Environment: deaths, injuries, hedonic losses due to life and community disruption. They also include lost wages in this category.

- Natural Environment: they include crop losses in this category.

They also discuss a number of considerations in risk evaluation and mitigation for the environments we want to protect.

J. Haas, R. Kates, M. Bowden, eds. 1977. *Reconstruction following disaster*. Cambridge Mass.: MIT Press.

This is a classic work on the subject of recovery from natural disaster. They conducted case studies of several major disasters, ranging from the Great San Francisco Earthquake to the Managua (Nicaragua) earthquake of 1972.

Much of the subsequent work on recovery uses the framework established by Haas et al. However, many of the researchers after Haas at al., disagree with the Haas framework. The problem is two-fold. (1) The Haas framework is more of a way of organizing and interpreting information than it is really an empirical result. (2) Subsequent researchers are interested in different questions. They are interested in differences in impact and recovery far more than they are interested in the degree of recovery. As Chang (2003) points out, "this more recent literature has been concerned with disparities and equalities in recovery, and with conceptualizing disaster recovery as a social process involving decision-making, institutional capacity, and conflicts between interest groups." So they are simply asking a different set of questions.

Kates and Pijawka (pp. 1-24) describe the model of recovery. Recovery is broken up into four different recovery periods: (1) the Emergency Period, (2) Restoration Period, (3) the Replacement Reconstruction Period, and (4) the Commemorative, Betterment, Developmental Reconstruction Period. They explicitly note that these periods may overlap, and different parts of the city may be in different stages at the same time.

They note that there is no *recent* example of a damaged/destroyed city not rebuilding. In a historical analysis covering several hundred years they could find no more than one or two cities which did not eventually recover their full populations after a disaster.

Haas et al. (pp. 25-68) discuss general observations about recovery, and the choices that public authorities make in guiding it. Disasters tend to accelerate suburbanization. They observe that disasters tend to accelerate existing trends in a city. So, in a city where business is shifting from one industry to another, a trend that might have taken several years will be completed in only a short time.

They observe that private actors want a prompt return to normal. The longer the government delays, the more likely it is that people will rebuild, with or without permission, and so the longer the planners take to make their decisions, the more likely it is that the decisions will be made for them.

Planners often have the idea that disasters serve as a form of "instant urban renewal," and represent an opportunity to rebuild the city so as to provide improved efficiency, equity, or amenity. Haas et al. observe that "despite the best efforts to shape the character of the reconstructed city, fundamental change is unlikely. Past trends will be accelerated in most cases." "Overambitious plans to accomplish [these] goals tend to be counterproductive." They also note that "a basic error of the professional community is to assume that formal studies…are requirements for reconstruction when there is already such a plan in the minds of the community inhabitants—the predisaster city."

Bowden et al. (pp. 69-145) discuss the details of the rebuilding of houses, businesses and neighborhoods. They are focused primarily on location.

In their analysis of San Francisco (1906) they note several details (the other case studies tended to support these observations):

- First, the wealthy reestablish housing first. The poor were several years in finding new houses, and many who lived there during the disaster left.

- Initially, with 50 % of the housing stock destroyed and the population not seriously depleted, housing prices were sky high. As time went on and the building stock was restored, prices fell.

- In San Francisco, the city expanded and reduced its density. The Central Business District grew in physical size, and residences moved further out.

- Sorting became much stronger. So rich and poor neighborhoods that had been somewhat intermixed before became strongly sorted. Ethnic enclaves became much sharper. The business district became much more sorted with the financial district cleanly separated from the garment district, from the theater district etc. Trends in terms of business types accelerated. So manufacturing was already moving out of San Francisco, and that trend accelerated.

- Naturally, small businesses were fairly likely to fail in the large disasters. That however does not address the question of how readily new businesses arose to take their place.

J. Harrald, 2006. "Agility and Discipline: Critical Success Factors for Disaster Response" *The Annals of the American Academy of Political and Social Science* 604 (1): 256-272.

For more than thirty years, the U.S. emergency management community has been increasing its ability to structure, control, and manage a large response. The result of this evolution is a National Response System based on the National Response Plan and the National Incident Management System.

Over the same period, social scientists and other disaster researchers have been documenting and describing the nonstructural factors such as improvisation, adaptability, and creativity that are critical to coordination, collaboration, and communication and to successful problem solving.

This article argues that these two streams of thought are not in opposition, but form orthogonal dimensions of discipline and agility that must both be achieved. The critical success factors that must be met to prepare for and respond to an extreme event are described, and an organizational typology is developed.

He divides response into four phases. (1) "The initial response is conducted by resources on the ground ... while external resources are mobilized." (2) "An integration phase is required to structure these resources into a functioning organization." (3) "A production phase is reached where the response organization is fully productive, delivering needed services as a matter of routine." (4) "Finally, the large external presence is diminished during a demobilization and transition to recovery stage."

K. Hosseini, M. Jafari, M. Hosseini, B. Mansouri, S. Hosseinioon, 2009. "Development of urban planning guidelines for improving emergency response capacities in seismic areas of Iran" *Disasters* 33 (4): 645-664.

This presents an evaluation of the disaster response to several large earthquakes in Iran between 1990 and 2006. The response and recommendations are based on experts' opinions as to the best ways to conduct emergency response.

The research concentrated on emergency response operations, emergency medical care, emergency transportation, and evacuation. They are primarily thinking of preplanning and prepositioning of resources so that when disaster strikes, everything is ready.

They propose portable medical triage units scattered throughout an impacted region. These have the ability to absorb casualties when they are high. Such units would deal with the lightly wounded without burdening hospitals and such, and stabilize the more severely wounded for transport to hospitals. In that vein, they note that 80 % of medical cases treated did not require hospitalization.

R. Iman, M. Johnson, C. Watson, 2005. "Sensitivity Analysis for Computer Model Projections of Hurricane Losses" *Risk Analysis* 25 (5): 1277-1297.

and

R. Iman, M. Johnson, C. Watson, 2005. "Uncertainty Analysis for Computer Model Projections of Hurricane Losses" *Risk Analysis* 25 (5): 1299-1312.

Florida requires disaster models used in justifying insurance rates be approved by the Florida Commission on Hurricane Loss Projection Methodology. As a part of the approval process the Commission requires that sensitivity and uncertainty analyses be conducted of the models. A sensitivity analysis would show how sensitive the results are to changes in specific input variables. An uncertainty analysis would show how much of the variance in output of the model is attributable to variance in specific input variables. These papers discuss methods for conducting these analyses.

S. King, A. Kiremidjian, N. Basöz, K. Law, M. Vucetic, M. Doroudian, R. Olson, J. Eidinger, K. Goettel, G. Horner, 1997. "Methodologies for Evaluating the Socio-Economic Consequences of Large Earthquakes" *Earthquake Spectra* 13 (4): 565-584.

> They developed a comprehensive methodology for evaluating the socio-economic impacts of large earthquakes. New models were developed for some of the methodology components, such as the identification and ranking of critical facilities. For other components, such as the estimation of building and lifeline component damage, existing models were adopted and modified for use within a GIS environment.
>
> The methodology was illustrated through a case study for the city of Palo Alto, California. Damage and loss estimates were made for four earthquake scenarios. Then peak ground acceleration, shaking intensity (interaction with soil types), and liquifaction probabilities were estimated. Then they estimated amount of damage, lives lost, and 'lifeline' damage.
>
> They came up with a list of 'critical facilities' based on the expected harm done. This used a weighting method to weight physical damage, lives lost, injuries, indirect effects etc. to determine the expected cost to society as a result of the earthquake. Indirect effects included things like fire damage resulting from a failure of the water supply system (fires not put out).
>
> This in turn fed into a benefit cost analysis to estimate the benefit of seismic retrofit.

Kunreuther, H., R. Meyer, and C. Van den Bulte. 2004. *Risk Analysis for Extreme Events: Economic Incentives for Reducing Future Losses*. NIST.

> They present a conceptual framework of how risk assessment, risk perception and risk management are linked with each other. Risk assessment evaluates the likelihood and consequences of prospective risks. Risk perception is concerned with the psychological and emotional aspects of risks. Risk management involves developing strategies for reducing the likelihood and/or consequences of extreme events.
>
> Regarding risk assessment, they describe how to generate an exceedance probability curve, which summarizes the risk and provides valuable input for different stakeholders to develop strategies for managing risk.
>
> Regarding risk perception they discuss how individual decisions on whether or not to adopt protective measures are influenced by psychological and emotional factors. They argue that individuals under-invest in mitigation and their explanations focus primarily on bounded rationality type explanations.
>
> Regarding risk management strategies, they focus on insurance and mitigation as two complementary strategies for reducing future losses and providing funds for recovery.

A major theme of the report is the need for public-private partnerships in managing disaster risk.

M. Lindell, R. Perry, 2000. "Household Adjustment to Earthquake Hazard: A Review of Research" *Environment and Behavior* 32 (4): 461-501.

They do a literature survey of 23 studies trying to understand what factors contribute to the adoption of earthquake mitigation measures by households. They find that households' adoption of earthquake hazard adjustments is correlated with:

- Perceptions of the hazard. That is, the greater the perceived hazard, the more likely it was that people would adopt mitigation measures.

- Alternative adjustments. That is, the less costly a mitigation measure was, the more alternative uses it had, and the more effective it was perceived to be, the more likely it is to be adopted.

- Social influences. Mitigation measures are more likely to be adopted if others around are adopting them.

- Demographic characteristics. Demographic factors do not correlate well with mitigation.

M. Lindell, C. Prater, 2000. "Household adoption of seismic hazard adjustments: A comparison of residents in two states" *International Journal of Mass Emergencies and Disasters* 18 (2): 317-338.

Residents of a high seismic hazard area were compared with those in a moderate seismic hazard area in terms of demographic characteristics, personal hazard experience, risk perception, hazard intrusiveness, and self-reported adoption of 16 hazard adjustments.

The results show that the two locations differed substantially in hazard experience, somewhat less so in risk perceptions and hazard intrusiveness, and little in hazard adjustment. Multiple regression analyses supported a causal chain in which location and demographic characteristics cause hazard experience, hazard experience causes hazard intrusiveness, perceived risk causes hazard intrusiveness, and hazard intrusiveness causes the adoption of hazard adjustments.

M. Lindell, D. Whitney, 2000. "Correlates of Household Seismic Hazard Adjustment Adoption" *Risk Analysis* 20 (1): 13-26.

This study examined the relationships of self-reported adoption of 12 seismic hazard adjustments with respondents' demographic characteristics, perceived risk, perceived hazard knowledge, perceived protection responsibility, and perceived attributes of the hazard adjustments.

Perceived attributes of the hazard adjustments differentiated among the adjustments and had stronger correlations with adoption than any of the other predictors. These results identify the adjustments and attributes that emergency managers should address to have the greatest impact on improving household adjustment to earthquake hazard.

Liu, Haibin, Rachel Davidson, and T. Apanasovich. 2007. "Statistical Forecasting of Electric Power Restoration Times in Hurricanes and Ice Storms." *Power Systems, IEEE Transactions on* 22:2270-2279.

The paper develops an empirical method for estimating the number of outages and the amount of time required to restore power after ice storms and hurricanes.

S. Miles, S. Chang, 2003. *Urban disaster recovery: a framework and simulation model.* MCEER.

This paper concerns the modeling of urban recovery from earthquake disasters with the emphasis on simulating recovery processes rather than on estimating dollar losses. A conceptual framework of disaster recovery is proposed that is then implemented in a prototype simulation model. The model emphasizes the dynamic or temporal processes of recovery, simulates impacts at the individual agent level of analysis, and relates recovery across business, household, and lifeline infrastructure sectors.

The prototype model is then applied to an actual event, the 1995 Kobe earthquake. The Kobe prototype successfully modeled some aspects of the Kobe recovery, but showed some serious deficiencies in other areas.

D. Mileti, 1999. *Disasters by Design: A Reassessment of Natural Hazards in the United States.* Washington D.C.: Joseph Henry Press.

He argues that U.S. Disaster losses are increasing at an unsustainable rate, and will soon become unsupportable. These losses, and the fact that there seems to be an inability to reduce such losses, are the consequences of narrow and short-sighted development patterns, cultural premises, and attitudes toward the natural environment, science, and technology. This is a problem that can be fixed by the application of sustainable community planning.

His principles for sustainable hazard mitigation are: (1) maintain and if possible enhance environmental quality; (2) maintain and if possible enhance people's quality of life; (3) foster local resiliency to and responsibility for disasters; (4) recognize that sustainable, vital local economies are essential; (5) identify and ensure inter- and intra-generational equity; and (6) adopt a consensus-building approach, starting with the local level.

The main solutions he advocates are:

- Get out of the way: don't build in the flood plain, or on the beach in hurricane country.

- Protect the environment.

- Good Building codes properly enforced.

- Hazard prediction and warning (in particular, some sort of national warning system).

D. Mileti, J. Sorensen, 1990. *Communication of emergency public warnings: a social science perspective and state-of-the-art assessment*. Oak Ridge National Laboratory.

More than 200 studies of warning systems and warning response were reviewed for this social-science perspective and state-of-the-art assessment of communication of emergency public warnings.

First, they discuss many current myths about public response to emergency warning that are at odds with knowledge derived from field investigations. Some of these myths include the "keep it simple" notion, the "cry wolf" syndrome, public panic and hysteria, and those concerning public willingness to respond to warnings.

Variations in the nature and content of warnings have a large impact on whether or not the public heeds the warning. Relevant factors include the warning source; the warning channel; the consistency, credibility, accuracy, and understandability of the message; and the warning frequency.

People want information from multiple sources, and respond to warnings by seeking more information. They don't blindly do what they are told. However, they may be persuadable. As a result, they recommend providing plenty of detail about the hazard. This helps people understand the risks, and the reasons for the recommended responses.

An important point they make is that the response to the warning needs to be monitored. If people react to the warning in ways that increase the hazard, adjustments to the warning need to be made.

Characteristics of the population receiving the warning affect warning response. These include social characteristics such as gender, ethnicity and age, social setting characteristics such as stage of life or family context, psychological characteristics such as fatalism or risk perception, and knowledge characteristics such as experience or training.

Finally, different methods of warning the public are not equally effective at providing an alert and notification in different physical and social settings. Most systems can provide a warning given three or more hours of available warning time. Special systems such as tone-alert radios are needed to provide rapid warning.

S. Miller, R. Muir-Wood, A. Boissonnade, 2008. "An exploration of trends in normalized weather-related catastrophe losses." pp. 225-247. In Diaz, Henry, and Richard Murnane, eds. *Climate extremes and society*. Cambridge University Press.

> They attempt to evaluate potential trends in global natural catastrophe losses, and estimate the impact of climate change on those trends. In doing so, they compensate for changes in asset values and exposures over time. They create a Global Normalized Catastrophe Catalogue covering weather-related catastrophe losses in the principal developed (Australia, Canada, Europe, Japan, South Korea, United States) and developing (Caribbean, Central America, China, India, the Philippines) regions of the world. They survey losses from 1950 through 2005, although data availability means that for many regions the record is incomplete for the period before the 1970's even for the largest events.
>
> After 1970, when the global record becomes more comprehensive, they find evidence of an annual upward trend for normalized losses of 2 % per year. Conclusions are heavily weighted by U.S. losses, and their removal eliminates any statistically significant trend. Large events, such as Hurricane Katrina and China flood losses in the 1990s, also exert a strong impact on trend results. In addition, once national losses are further normalized relative to per capita wealth, the significance of the post-1970 global trend disappears. They find insufficient evidence to claim a statistical relationship between global temperature increase and normalized catastrophe losses.

A. Muermann, H. Kunreuther, 2008. "Self-protection and insurance with interdependencies" *Journal of Risk and Uncertainty* 36 (2): 103-123.

> They study optimal investment in self-protection of insured individuals when they face interdependencies in the form of potential externalities from others. If individuals cannot coordinate their actions, then the positive externality of investing in self-protection implies that, in equilibrium, individuals under-invest in self-protection. Limiting insurance coverage through deductibles or selling "at-fault" insurance can partially internalize this externality and thereby improve individual and social welfare.

National Research Council, 1999. *The Impacts of Natural Disasters: A Framework for Loss Estimation* Washington DC: National Academies Press.

> The committee compiled a list of sources for disaster losses. They recommend that a national database of losses be compiled, and recommend Bureau of Economic Analysis as the lead agency. Data should be collected for all "major" disasters, which they define as anything with total damage over $25 Million or if so declared by the President under the Stafford Act.

They note that there will be demand for early estimates, but that final totals will not be known for many months (or even years). So they suggest that the estimates be subject to revisions as new data come in, much like routine economic data.

They suggest that the database include (among other things):

- Physical damage:

- Event

- Type of disaster (hurricane, flood, earthquake, etc.)

- Location - ideally to county or ZIP-code level

F. Norris, M. Friedman, P. Watson, C. Byrne, E. Diaz, K. Kaniasty, 2002. "60,000 Disaster Victims Speak: Part I. An Empirical Review of the Empirical Literature, 1981–2001" *Psychiatry: Interpersonal & Biological Processes* 65 (3): 207-239.

> Results for 160 samples of disaster victims were coded as to sample type, disaster type, disaster location, outcomes and risk factors observed, and overall severity of impairment.
>
> Regression analyses showed that samples were more likely to have high impairment if they were composed of youth rather than adults, were from developing rather than developed countries, or experienced mass violence (e.g., terrorism, shooting sprees) rather than natural or technological disasters. Most samples of rescue and recovery workers showed remarkable resilience. Within adult samples, more severe exposure, female gender, middle age, ethnic minority status, secondary stressors, prior psychiatric problems, and weak or deteriorating psychosocial resources most consistently increased the likelihood of adverse outcomes.
>
> It seems likely that a substantial portion of the differences between first and third world effects is a product of differential sampling. Sampling issues pervade the paper. This is not a random sample of disasters, even within the U.S..

F. Norris, S. Stevens, B. Pfefferbaum, K. Wyche, R. Pfefferbaum, 2008. "Community Resilience as a Metaphor, Theory, Set of Capacities, and Strategy for Disaster Readiness" *American Journal of Community Psychology* 41 (1-2): 127-150.

> They draw upon literatures in several disciplines, to present a theory of resilience that encompasses contemporary understandings of stress, adaptation, wellness, and resource dynamics. Community resilience is a process linking a network of adaptive capacities (resources with dynamic attributes) to adaptation after a disturbance or adversity.
>
> They say "across these definitions, there is general consensus on two important points: first, resilience is better conceptualized as an ability or process than as an outcome…; and

second, resilience is better conceptualized as adaptability than as stability…. In fact, in some circumstances, stability (or failure to change) could point to lack of resilience."

Community resilience emerges from four primary sets of adaptive capacities—Economic Development, Social Capital, Information and Communication, and Community Competence—that together provide a strategy for disaster readiness.

To build collective resilience, communities must reduce risk and resource inequities, engage local people in mitigation, create organizational linkages, boost and protect social supports, and plan for not having a plan, which requires flexibility, decision-making skills, and trusted sources of information that function in the face of unknowns.

R. Pielke, M. Downton, 2000. "Precipitation and Damaging Floods: Trends in the United States, 1932–97" *Journal of Climate* 13 (20): 3625-3637.

They examine the time trend for flooding. As usual, damage from floods is increasing exponentially. After accounting for population growth it is still increasing exponentially, but after accounting for increase in National wealth, flood damage is decreasing.

Flood damage is positively correlated with extreme precipitation. They also show that extreme precipitation is increasing over time. So climate change is showing up in the damage record. However damage per unit of wealth is decreasing with time. So we must be becoming less vulnerable to flood damage with time per unit of wealth.

R. Pielke, C. Landsea, 1998. "Normalized Hurricane Damages in the United States: 1925–95" *Weather and Forecasting* 13 (3): 621-631.

Previous research into long-term trends in hurricane-caused damage along the U.S. coast has suggested that damage has been quickly increasing within the last two decades, even after considering inflation. However, to best capture the year-to-year variability in tropical cyclone damage, consideration must also be given toward two additional factors: coastal population changes and changes in wealth.

Both population and wealth have increased dramatically over the last several decades and act to enhance the recent hurricane damages preferentially over those occurring previously. More appropriate trends in the United States hurricane damages can be calculated when a normalization of the damages are done to take into account inflation and changes in coastal population and wealth.

With this normalization, the trend of increasing damage amounts in recent decades disappears. Instead, substantial multi-decadal variations in normalized damages are observed: the 1970s and 1980s actually incurred less damages than in the preceding few decades. Only during the early 1990s does damage approach the high level of impact seen back in the 1940s through the 1960s, showing that what has been observed recently is not unprecedented.

Over the long term, the average annual impact of damages in the continental United States is about $4.8 billion (1995 $), substantially more than previous estimates. Of these damages, over 83 % are accounted for by the intense hurricanes (Saffir–Simpson categories 3, 4, and 5), yet these make up only 21 % of the U.S.-landfalling tropical cyclones.

Poland, Chris, David Bonowitz, Joe Maffei, and Christopher Barkley. 2009. "The Resilient City." *Urbanist* 4-21.

This is a plan for disaster mitigation, response and recovery for the City of San Francisco. The report includes targets for service levels, recommendations for building code modifications and retrofit requirements, recommendations for upgrading lifeline systems, emergency response and recovery.

E. Quarantelli, 1983. *Delivery of emergency medical services in disasters: assumptions and realities*. New York N.Y.: Irvington Publishers.

Emergency response in disasters is poorly coordinated (if at all) with very little communication between components. Plans rarely work as intended. Partially, this is because much of the response is by people who are normally outside the EMS system. However, even within the normal EMS system there is little coordination between (say) police and fire departments.

Fire and police are usually in control of the site and in charge of search and rescue. However, in search and rescue, much is done by bystanders, and there is very little real control or coordination. A lot of people and resources self-respond. That can overwhelm some aspects of response.

There is little communications between the site and the hospital(s). Hospitals get most of their information from patients and ambulance drivers. Initial guesses as to casualties are generally wildly high.

Very little triage at the site occurs. In general there is a tendency to transport victims as quickly as possible to the hospital. As a practical matter, that means the less injured get transported before the critically injured. Transportation broke down by approximately 60 % ambulance, 24 % private vehicle (including foot), 16 % police car. In large disasters, a significant percentage of transporters are from outside the area.

Sometimes there are more patients than ambulances. However, this will only apply to very large scale disasters. The usual problem is that there are more ambulances than patients.

The number of patients can strain emergency rooms. This is especially true since patients are not distributed evenly to hospitals in the area. In mass disasters, hospitals just do not keep good records of patients, injuries, and treatments. Especially for those with minor injuries.

The few suggestions of inadequate treatment mostly come from overresponse. That is resources are pulled to the disaster away from other critical needs. So, ambulance response leaves other areas temporarily unserved (or underserved), and hospital personnel may be pulled away from critical care areas leaving other patients uncared for.

E. Quarantelli, 1982. *Sheltering and housing after major community disasters: case studies and general observations*. Disaster Research Center: University of Delaware.

He divides shelter into four categories:

- Emergency Sheltering: Victims seeking shelter outside their own homes for short periods of time (hours).

- Temporary Sheltering: Victims seeking shelter outside their own homes for the duration of the emergency (usually a few days). Household routines are not reestablished. Here, the issue of how the sheltered will be fed becomes an issue, where in the former case it usually does not.

- Temporary Housing: Household routines are reestablished.

- Permanent Housing

A. Rose, S. Liao, 2005. "Modeling Regional Economic Resilience to Disasters: A Computable General Equilibrium Analysis of Water Service Disruptions" *Journal of Regional Science* 45 (1): 75-112.

Recent natural and manmade disasters have had significant regional economic impacts. These effects have been muted, however, by what they term 'resilience,' which is the inherent ability and adaptive response that enable firms and regions to avoid potential losses. In this category of 'resilience' they includes things like input and import substitution, use of inventories, conservation, the rationing feature of markets, and the ability to import critical inputs from other regions.

Computable general equilibrium (CGE) analysis is a promising approach to disaster impact analysis because it is able to model the behavioral response to input shortages and changing market conditions. However, without further refinement, CGE models, as well as nearly all other economic models, reflect only "business-as-usual" conditions, when they are based on historical data. This paper advances the CGE analysis of major supply disruptions of critical inputs by: specifying operational definitions of individual business and regional macroeconomic resilience, linking production function parameters to various types of producer adaptations in emergencies, developing algorithms for recalibrating production functions to empirical or simulation data, and decomposing partial and general equilibrium responses. They illustrate some of these contributions in a case study of the sectoral and regional economic impacts of a disruption to the Portland Metropolitan Water System in the aftermath of a major earthquake.

A. Rose, D. Lim, 2002. "Business interruption losses from natural hazards: conceptual and methodological issues in the case of the Northridge earthquake" *Global Environmental Change Part B: Environmental Hazards* 4 (1): 1-14.

> The paper contains a valuable discussion on the problems involved in counting business interruption costs. In particular, since the value of a business asset is the discounted value of its income, counting both replacement value and business interruption costs involves some double-counting.
>
> They estimate the business interruption losses during the Northridge Earthquake due to power outages in the LADWP service area. They adjust for 'resiliency:' that is the ability of firms to adjust so as to minimize losses.
>
> What they do not do is adjust for other interruptions. Just because power is restored does not mean a firm is operating. One way to correct for that would be to use power deliveries to measure business operation.

A. Rose, G. Oladosu, S. Liao, 2007. "Business Interruption Impacts of a Terrorist Attack on the Electric Power System of Los Angeles: Customer Resilience to a Total Blackout" *Risk Analysis* 27 (3): 513-531.

> They estimate the largest category of economic losses from electricity outages—business interruption—in the context of a total blackout of electricity in Los Angeles. In particular, they do so while taking into account indirect effects and resilience.
>
> When a disaster occurs effects occur at three different degrees of separation:
>
> - First, you have those businesses, structures, and utilities that are directly damaged.
>
> - Second, there are businesses that cannot operate because a critical utility is off. Firms that cannot operate because the power (or water) is out, or because their workers cannot get to work due to road outages fall into this category.
>
> - Third, there are businesses that cannot operate because a key supplier or customer is not operating. So, a factory 200 miles away from the disaster that cannot operate because its parts supplier is down falls into this category.
>
> They consider only the third category to be 'indirect costs.' The third category is similar to the 'multiplier effect' from macroeconomics.
>
> The results indicate that indirect effects in the context of general equilibrium analysis are moderate in size. The stronger factor, and one that pushes in the opposite direction, is resilience. Their analysis indicates that electricity customers have the ability to mute the potential shock to their business operations by as much as 86 %. Moreover, market resilience lowers the losses, in part through the dampening of general equilibrium effects.

A. Rose, 2004. "Economic Principles, Issues, and Research Priorities in Hazard Loss Estimation." pp. 13-36. In Okuyama, Yasuhide, and Stephanie Chang, eds. *Modeling Spatial and Economic Impacts of Disasters*. Springer.

>He reviews the relative strengths and weaknesses of Input-Output analysis and Computable General Equilibrium analysis for the estimation of business interruption losses in disasters.
>
>Input-Output analysis is inflexible and linear. So firms in an IO model do not find alternate sources of inputs when a supplier goes down. There are no increasing or decreasing returns to scale.
>
>Computable General Equilibrium analysis assumes equilibrium and instantaneous adjustment. So, firms find new suppliers instantly, prices adjust to the new supply-demand schedule instantly, etc.
>
>He recommends that "I-O models, with an adjustment for inventories, are probably better suited to recovery periods of less than one week, but CGE models are better suited to all other cases...."
>
>He talks briefly about econometric estimation of macroeconomic models. He says this has been only rarely used due to the data requirements of such a model.

Rose, Adam et al. 2007. "Benefit-Cost Analysis of FEMA Hazard Mitigation Grants." *Natural Hazards Review* 8:97-111.

>This paper reports on a study that applied Benefit-Cost Analysis methodologies to a statistical sample of the nearly 5,500 Federal Emergency Management Agency (FEMA) mitigation grants between 1993 and 2003 for earthquake, flood, and wind hazards. HAZUS-MH was employed to assess the benefits, with and without FEMA mitigation in regions across the country, for a variety of hazards with different probabilities and severities. The results indicate that the overall benefit-cost ratio for FEMA mitigation grants is about 4:1, though the ratio varies from 1.5 for earthquake mitigation to 5.1 for flood mitigation. Sensitivity analysis was conducted and shows these estimates to be quite robust.

C. Rubin, 1985. *Community Recovery from a Major Natural Disaster*. Institute of Behavioral Science, University of Colorado.

>She conducts a series of case studies of disasters to see how they handled recovery. None of her disasters were particularly large. In terms of response, it seems unlikely that any of them seriously strained the resources available to the local communities.
>
>She suggests three factors that matter to recovery: Leadership, Ability to Act, and Knowledge. She also suggests a number of propositions leading to good recovery.

One implicit lesson on these comments is either have someone with a clear vision who can carry through their vision quickly, or have a preexisting community consensus as to what the recovered community should look like.

D. Stephenson, 2008. "Definition, diagnosis, and origin of extreme weather and climate events." pp. 11-23. In Diaz, Henry, and Richard Murnane, eds. *Climate extremes and society*. Cambridge University Press.

Extreme weather and climate events are a major source of risk for all human societies. Various societal changes, such as increased populations in coastal and urban areas and increasingly complex infrastructure, have made us potentially more vulnerable to such events than we were in the past. In addition, the properties of extreme weather and climate events are likely to change in the twenty-first century owing to anthropogenic climate change.

The definition, classification, and diagnosis of extreme events are far from simple. There is no universal unique definition of what is an extreme event. He notes that an 'extreme' weather event (defined as one that occurs rarely) is not the same as a 'high-impact event.' He develops a working definitions of an extreme weather event and evaluates their rate of occurrence through time. He finds that 'rare' flood events have been increasing in frequency in the UK.

H. Tatano, K. Yamaguchi, N. Okada, 2004. "Risk Perception, Location Choice and Land-use Patterns under Disaster Risk: Long-term Consequences of Information Provision in a Spatial Economy." pp. 163-177. In Okuyama, Yasuhide, and Stephanie Chang, eds. *Modeling Spatial and Economic Impacts of Disasters*. Springer.

They develop a model of choice where people hear what experts have to say about a risk, and act in ways different from what the experts recommend. In their information model, people have (uninformative) beliefs about a hazard, and are given expert information about location hazard that they consider noisy (but which is assumed in this paper to be perfectly reliable). So how far their belief is from the 'true' value of the hazard depends on how reliable they consider the expert information relative to their initial beliefs.

They then derives housing prices as a function of commuting distance, exposure to hazard, and beliefs about the reliability of the expert information.

The Infrastructure Security Partnership, 2006. *Regional disaster resilience: a guide for developing an action plan*. Reston Va.: American Society of Civil Engineers.

They define disaster resilience as the capability to prevent or protect against significant multihazard threats and incidents, including terrorist attacks, and to expeditiously recover and reconstitute critical services with minimum damage to public safety and health, the economy, and national security.

They say that "comprehensive regional preparedness is key to ensuring that communities, states, and the nation can expeditiously respond to and recover from disasters of all types, particularly extreme events," and develop a list of resources and planning guidelines for preparing for a disaster.

Thomas, Douglas, and Robert Chapman. 2008. *A Guide to Printed and Electronic Resources for Developing a Cost-Effective Risk Mitigation Plan for New and Existing Constructed Facilities*. NIST SP 1082. Gaithersburg, MD: NIST.

This paper is provides a list of resources useful for implementing the ASTM 2506 protocol for evaluating the cost effectiveness of proposed disaster mitigation measures. This document provides an annotated bibliography of printed and electronic resources that serves as that central source of data and tools to help the owners, managers, and designers of constructed facilities develop a cost-effective risk mitigation plan.

K. Tierney, M. Lindell, R. Perry, 2001. *Facing the unexpected: disaster preparedness and response in the United States*. Washington D.C.: Joseph Henry Press.

This is a summary of the existing literature. Among other findings, they observe that there is little or no empirical research on adoption of mitigation measures by governments. 'Preparedness' is also poorly correlated (or perhaps even uncorrelated) with response effectiveness. In any particular disaster it is difficult to determine whether results are attributable to planning and preparation, luck, or good improvisation.

K. Tierney, 2006. "Social Inequality, Hazards, and Disasters." pp. 109-128. In Daniels, Ronald, Donald Kettl, and Howard Kunreuther, eds. *On risk and disaster: lessons from Hurricane Katrina*. Philadelphia: University of Pennsylvania Press.

Poverty, lack of education, disability, race, etc. increase vulnerability. Poverty primarily works through budget constraints, and can constrain (among other things) the ability to evacuate. It can also affect the cost of evacuation, in the sense that poor, hourly workers are more likely to lose income, and larger percentages of their income, if they evacuate. In the U.S., the assumption is that people have transportation to manage their own evacuation. In New Orleans, that was not the case, and she implies that that often holds for the poor.

In general, she suggests that disaster preparedness and assistance programs are not all that well tailored for the poor and disadvantaged. This includes, among other things, that richer victims are also better able to search for and find assistance.

One of her themes is that several of these characteristics impact behavior and needs more or less independent of vulnerability. So the disabled are specifically impaired from evacuating when necessary. Some ethnic groups preferentially have multi-family living arrangements or extended family living arrangements, and that impacts the kind of aid they need.

K. Tierney, 1997. "Impacts of Recent Disasters on Businesses: The 1993 Midwest Floods and the 1994 Northridge Earthquake." pp. 189-222. In Jones, Barclay, ed. *Economic consequences of earthquakes: Preparing for the unexpected.* NCEER.

> This paper presents findings from two Disaster Research Center surveys on disaster-related business impacts. The first study, conducted in 1994, focuses on the ways in which the 1993 Midwest floods affected the operations of businesses in Des Moines / Polk County, Iowa. The second project uses a similar methodological approach to assess the impacts of the 1994 Northridge earthquake on businesses in Los Angeles and Santa Monica, CA. Both studies utilize large representative samples that include both large and small firms and a range of business types.
>
> Previous studies found that marginal businesses were hardest hit, as well as those that leased (rather than owned) their location, lost expensive inventory, or were heavily dependent of foot traffic and had to relocate. Smaller businesses and those in services were more vulnerable. Indirect losses resulted from damage to transportation system, problems with employee or customer access, and shipping delays.
>
> She found that most businesses self-insured. Most businesses were about as well off as before the disaster. Of the rest, they were about equally split between those better off and those worse off. Indirect effects were a big part of losses, especially in Des Moines where most businesses were not directly affected by flooding. 'Lifeline' interruptions were a major cause of losses.

K. Tierney, 1997. "Business impacts of the Northridge earthquake" *Journal of Contingencies and Crisis Management* 5 (2): 87-97.

> This paper focuses on the immediate and longer-term impacts the earthquake had on businesses in the Greater Los Angeles region after the Northridge earthquake. The data reported here are based on a survey that the Disaster Research Center, at the University of Delaware, conducted with a representative, randomly-selected sample of businesses in the cities of Los Angeles and Santa Monica, two jurisdictions that were particularly hard-hit by the earthquake.
>
> The paper addresses the following research questions: (1) What direct impacts and losses did businesses experience in the earthquake? (2) In what ways did the earthquake affect the operations of the businesses studied? If they experienced business interruption, why were they forced to close? What other kinds of problems did business have to cope with following the earthquake? (3) What earthquake preparedness measures had businesses undertaken prior to the disaster, and what have they done subsequently to prepare? and (4) To what extent have business operations returned to pre-earthquake levels, and which businesses appear to be experiencing the most difficulty with recovery?
>
> She found that about 2/3 of businesses reporting damage reported non-structural damage. The median loss was about $5,000, but the mean loss was about $156,000. As usual, the

damage is highly skewed. About 13 % of businesses reported that their building had been declared 'unsafe'. Average closure time was approximately 2 days. Reasons include 'lifeline' interruptions and cleanup.

U.S. Government Accountability Office. 2007. *NATURAL DISASTERS: Public Policy Options for Changing the Federal Role in Natural Catastrophe Insurance*. Washington DC: U.S. GAO.

The purpose of the document is to explore ways of restructuring the 'Catastrophe' Insurance market for the following objectives:

- Charging premium rates that fully reflect actual risks,

- Encouraging private markets to provide natural catastrophe insurance,

- Encouraging broad participation in natural catastrophe insurance programs, and

- Limiting costs to taxpayers before and after a disaster.

The short version is that these are probably incompatible, especially after taking into account the Public Choice aspects of the market.

J. Vigdor, 2007. "The Katrina Effect: Was There a Bright Side to the Evacuation of Greater New Orleans?" *The B.E. Journal of Economic Analysis & Policy: Advances* 7 (1): Article 64.

He uses longitudinal data from Current Population Surveys conducted between 2004 and 2006 to estimate the net impact of Hurricane Katrina-related evacuation on various indicators of well-being. While evacuees who have returned to the affected region show evidence of returning to normalcy in terms of labor supply and earnings, those who persisted in other locations exhibit large and persistent gaps, even relative to the poor outcomes of individuals destined to become evacuees observed prior to Katrina.

Overall, there is little evidence to support the notion that poor underemployed residents of the New Orleans area were disadvantaged by their location in a relatively depressed region. Evacuees started out worse than people who never left. People who never returned started out worse than people who left and came back. People who never returned did worse (on average, of course) than they did before they left.

J. Vigdor, 2008. "The economic aftermath of Hurricane Katrina" *The Journal of Economic Perspectives* 22 (4): 135–154.

He notes that it is the norm for cities to recovery. The only Pre-New Orleans, modern example he can find where a city did not was Dresden. That includes a number of bombed-out cities in WWII, including Hiroshima, Nagasaki and Warsaw.

He argues (paralleling Haas) that the only cities that do not recover from disaster are those already in decline. So, as with Haas, the disaster serves to accelerate a process already ongoing. Since (he argues) New Orleans has been in a long-term economic decline, he does not expect it to fully bounce back.

Viscusi, W. K. 2009. "Valuing risks of death from terrorism and natural disasters." *Journal of Risk and Uncertainty* 38:191-213.

He conducts a survey where he essentially asks people whether they value disaster deaths more or less than traffic deaths. The basic question he asks of survey participants can be paraphrased as "Would you rather spend your money to prevent x traffic deaths or y natural disaster deaths."

He found that natural disaster deaths are worth only half of what traffic deaths are worth. Terrorism deaths are worth about the same as traffic deaths.

Risk exposure is (very) weakly related to preferences on disasters, via income. It is strongly related in regard to terrorism. Risk aversion is weakly related to several of these via Seat-Belt Usage.

Waugh, W., and K. Tierney, eds. 2007. *Emergency management: principles and practice for local government*. 2nd ed. Washington D.C.: ICMA Press.

This is a textbook on emergency management and response to disasters. Its intended audience is local government officials who are tasked with developing and implementing emergency response plans and managing the local response to disasters.

G. Webb, K. Tierney, J. Dahlhamer, 2000. "Businesses and Disasters: Empirical Patterns and Unanswered Questions" *Natural Hazards Review* 1 (2): 83-90.

Through five systematic, large-scale mail surveys conducted since 1993, the Disaster Research Center obtained data on hazard awareness, preparedness, disaster impacts, and short- and long-term recovery among 5,000 private-sector firms in communities across the United States (Memphis/Shelby County, Tenn.; Des Moines, Iowa; Los Angeles, Calif.; Santa Cruz County, Calif.; and South Dade County, Fla.).

This paper summarizes findings from those studies in three major areas: (1) factors influencing business disaster preparedness; (2) disaster-related sources of business disruption and financial loss; and (3) factors that affect the ability of businesses to recover following major disaster events.

Business 'Preparedness' is operationally defined as fulfilling a checklist prepared by the researchers. By that definition, Business Preparedness is pretty thin. Bigger businesses are more likely to adopt mitigation measures. Measures that enhance life-safety are more likely to be adopted than measures that enhance business continuity and survival. As usual, cost and benefit matter to the choices in implementation. So, the cheaper a measure is, the more likely to be adopted; and the more alternative uses it has the more likely to be adopted.

Business interruption is most often (in the disasters reviewed) a product of 'lifeline' damage—electricity, water, communication, and transportation interruptions.

Bigger businesses do better in recovery. Preparedness has no relation to recovery. Construction firms are better off. Businesses in declining industries are less likely to recover than businesses in growing industries. Aid does not seem to have any impact on recovery.

They suggest that the reason why preparedness has so little to do with recovery is that (among other reasons) the preparedness measures recommended and implemented are aimed at enhancing response, not recovery.

D. Wenger, E. Quarantelli, R. Dynes, 1989. *Disaster analysis: Police and fire departments.* Disaster Research Center: University of Delaware.

Both police and fire departments tended to avoid tasks different than those related to their traditional roles. So, fire departments engaged in search and rescue, put out the fires, and then left. Police departments engaged in limited search and rescue, and did traffic, site, and crowd control.

Traffic and Crowd Control were difficult and not very successful in any of the cases. Congestion at the site was a problem in most of the cases they looked at. In some cases, needed emergency personnel and equipment could not be gotten to the site due to congestion.

Interagency coordination is always (and not just in these case studies) an issue. In effect, though, the lack of formal coordination between police and fire was rarely a problem. There was usually an implicit informal coordination taking place, where they ceded to each other their respective expertises.

The authors were generally not impressed with the Incident Command System where it appeared in these cases. It in effect ended up being a Fire Management System, and did a poor job of coordinating multiple agencies and the numerous volunteers who were present for several of the disasters. It did not even do a very good job of coordinating different fire organizations. The tendency to bump the Incident Commander task up the chain as soon as a new officer came on the scene was something they considered a potential problem.

D. Wenger, E. Quarantelli, R. Dynes, 1986. *Disaster Analysis: Emergency Management Offices And Arrangements.* Disaster Research Center: University of Delaware.

Numerous previous cases studies had found that planning and preparedness were deficient and that communications were a major issue in disaster response. Adequate resources generally were available; the problem was one of distributing them. They found in these case studies that planning and preparedness had improved. There was some evidence that it had brought some improvement to disaster response. However, planning did not guarantee effective response.

> The key to effective Local Emergency Management Agencies really was extensive previous experience. The areas of most difficulty in their case studies were communication, task assignment and coordination, and authority relationships. Essentially, you had independent units working, not talking to each other, and no one (that everyone agreed upon) in charge. Having a Local Emergency Management Agency did not eliminate this problem. However, planning does appear to ease these problems.

C. West, D. Lenze, 1994. "Modeling the Regional Impact of Natural Disaster and Recovery: A General Framework and an Application to Hurricane Andrew" *International Regional Science Review* 17 (2): 121-150.

> There are a number of approaches to modeling and estimating the impacts of natural disasters. However, many of those approaches require a great deal of specific detailed information to answer important questions regarding the impacts. This is a careful evaluation of the specific impacts of Hurricane Andrew on south Florida, intended for use in impact modeling and estimation.
>
> They use a wide variety of data sources, and pick the impacts apart piece by piece. They estimate physical damage by aggregating information from a variety of sources (all of which are incomplete and of inconsistent quality), labor market adjustments, and attempted to estimate demand and population impacts. So, for example, they estimate the number of people to leave the area based on specific known job losses (like the closure of Homestead AFB) and other factors.
>
> They also discuss challenges facing most regional models in estimating disaster impacts. In particular, most regional models do not explicitly account for capital stock (which is where the major direct impact of a disaster falls), and do not have a fully specified housing market. Level of aggregation for modeled sectors can be an issue, especially when the demand induced by the disaster within a sector has different characteristics from average demand absent a disaster.

J. Wright, P. Rossi, S. Wright, E. Weber-Burdin, 1979. *After the clean-up: Long-range effects of natural disasters*. Beverly Hills, Calif: Sage.

> They are interested in the long-range effects of disasters on communities. They analyze all disasters to hit the United States in the 1960's, by county and census tract. Based on their database, they could find no evidence of long-term impact from a disaster on either the county or census tract it was in.
>
> Note that the overwhelming majority of the disasters are small: the median tornado, for example, killed no one and caused less than $500,000 in damage. So, the impact of really large events may be obscured by a forest of small events.

Appendix B   Selected Databases, Software Tools, and Web Portals

The data sources listed below fall into three broad categories. There are several data sources that provide data on disaster damages. These include CRED, flooddamagedata.org, SHELDUS, Swiss Re, and the Property Claims Service. Second, there are data sources on hazards, including NOAA, NIFC and the U.S.GS. The remaining data sources provide macroeconomic, demographic or geographic data that are useful for analyzing disaster loss data.

The disaster loss data sources are not independent. They all draw from the same raw sources for their estimates of damages. However, they have different conventions for including the data in the database, and different scope. In addition, scope, accuracy, and methodologies have changed over time, so results on time trends may not be reliable.

Software includes several "Catastrophe" models (including HAZUS-MH, RMS Inc., and EQECAT), and a number of software packages designed for the development of regional macroeconomic models.

Data Sources

American Chamber of Commerce Research Association,
ACCRA Cost Of Living Index
http://www.coli.org/.

They maintain regional Cost of Living Indices.

BEA, U.S. Department of Commerce

The Bureau of Economic Analysis collects and publishes data for the U.S. Economy. In particular, the following data source is of interest for this research effort.

Fixed Asset Tables
http://www.bea.gov/national/FA2004/index.asp.

The Fixed Asset tables record the constructed building stock, and other fixed and durable assets, by year. The data goes back to 1925. Data are annual, and are broken down by asset category, industry, and legal form of organization.

Center for Research on the Epidemiology of Disasters,
CRED Website
http://www.cred.be/.

They are a division of the World Health Organization, and collect data on disaster losses (and deaths) world-wide. A preliminary review of the data suggests that it is not consistent with similar data collected for the U.S. from other sources. "The database is compiled from various sources, including UN agencies, non-governmental organizations, insurance companies, research institutes and press agencies. Priority is given to data from UN agencies, governments and the International Federation of Red Cross and Red Crescent Societies."

T. Chandler, 1987. *Four thousand years of urban growth: an historical census*. Lewiston N.Y. U.S.A.: St. David's University Press.

A compilation of population estimates for the world's largest cities over a period of approximately 4,000 years. The data are fairly low-resolution. So for the earlier periods, there may be population estimates only every century at best. The data get better as we move closer to the present.

Flood Damage in the United States
http://www.flooddamagedata.org/.

Flood damage data have been collected and published by the National Weather Bureau since 1933, with a hiatus between 1980 and 1982. Downton et al. (2005) compiled and published that data, and reanalyzed it for scope accuracy and consistency. The data set has problems, some of which are discussed below, but they argue that it still represents a good, long-term data set on flood damage.

Some of the characteristics of the data they observe are:

- The original data for the most part were estimated to the level of order of magnitude (e.g., between $500,000 and $5 Million), were collected by personnel untrained in estimating damages, were not checked for accuracy, and were incomplete.

- In general, however, the larger the event, the more accurate (in log terms) the estimate.

- The data are bottom censored. That means the totals are probably biased down.

- However, most of the damage occurs in a few very large events. So, for example, in Pennsylvania (which has the highest total damage in the database), something like 75 % of the total damage (over 75 years) occurred in a single flood event.

- With those caveats, they consider the data to be unbiased.

Hazards and Vulnerability Research Institute
Department of Geography
University of South Carolina
Spatial Hazard Events and Losses Database for the United States (SHELDUS)
http://webra.cas.sc.edu/hvri/products/sheldus.aspx.

This is an effort at a comprehensive national loss inventory of natural hazards. The database is described in detail in Cutter et al. (2008).The data starts in (approximately) 1960 and continues to the present. The data are drawn from a variety of mostly governmental sources. As such the data are not independent of the flood data at www.flooddamagedata.org. Several observations about the data are worth mentioning:

- Their loss estimates are conservative. That is, when a range of values are reported in their original sources they enter the lowest available estimate of damages in their database. That biases the damage data down. It also biases trends in the data. Since older data are less accurate and less precise, the downward bias becomes greater for the older data. So any attempt to find a trend will find one based solely on the biases in the dataset.

- Some types of disasters are underreported. They note in particular that drought is under reported.

- Of course, the data are bottom censored. The bottom censoring gets worse the older the data.

- There is no indexing by event. That can to some extent be remedied ex post, but only to some extent.

- Puerto Rico, Guam, and other U.S. territories are not included.

- Damage is distributed by county, often by simply dividing the estimated total damage by number of counties. So the spatial heterogeneity of damage is not reflected in the database.

ISO
Property Claims Services
http://www.iso.com/Products/Property-Claim-Services/Property-Claim-Services-PCS-info-on-losses-from-catastrophes.html.

PCS investigates reported disasters and determines the extent and type of damage, dates of occurrence, and geographic areas affected. To gather the information needed for its services, PCS has developed a national network of insurer claim departments, insurance adjusters, emergency managers, insurance agents, meteorologists, and fire and police officials. PCS also maintains contact with industry representatives in many countries.

Minnesota IMPLAN Group, Inc.
http://implan.com/v3/.

They are publishers of software, regional Social Accounting Matrices, and consulting services. The regional Social Accounting Matrices in particular are frequently used in the estimation of disaster impacts. An IMPLAN Social Accounting Matrix is at the heart of the FEMA hazard loss model, HAZUS-MH .

National Interagency Fire Center,
Fire Information - Wildland Fire Statistics
http://www.nifc.gov/fire_info/fire_stats.htm.

Statistics on the number and size of wildland fires in the U.S., including some estimates of losses.

National Oceanic and Atmospheric Administration
Atlantic basin hurricane database (HURDAT)
http://www.aoml.noaa.gov/hrd/hurdat/

This is a list of all known hurricanes in the Atlantic since 1851, with estimates of their magnitude. The data are described in Landsea et al. (2003). Naturally the earlier data are incomplete Storms that never made landfall before 1910 are likely not in the database. They estimate that 0-6 storms were missed per year before 1875 and 0-4 before 1910.

Some areas were sparsely populated, and so storms may have been missed even if they made landfall. For example, they figure that records for south Florida are unreliable before 1900, and for south Texas before 1880.

Oak Ridge National Laboratory,
LandScan Global Population Database
http://www.ornl.gov/sci/landscan/.

LandScan provides estimates of the population distribution worldwide down to the 1 km^2 scale.

Risk Management Solutions, Inc., *RMS Software*
http://www.rms.com/.

They have hazard and risk data for primarily the U.S., including flood zones, distance to coast, hurricane risk scores and profiles, distance to faults, earthquake risk scores, predicted ground shaking intensities for index quakes, soil types, liquefaction susceptibility, landslide susceptibility, Alquist-Priolo zones, convective storm risk scores, slope and elevation files.

They also have current insurance exposure data, and estimated insured loss data for predicted hazards.

Swiss Re, Inc.

Swiss Re is a reinsurance company that also collects and publishes data on insured losses from natural disasters.

U.S. Department of Housing and Urban Development,

Fair Market Rents
http://www.huduser.org/portal/datasets/fmr.html.

U.S. Geological Survey

The U.S.GS collects a variety of data regarding natural hazards and losses that would be useful in any evaluation of disasters and resilience. An incomplete list includes:

PAGER - Prompt Assessment of Global Earthquakes for Response
http://earthquake.usgs.gov/earthquakes/pager/.

Quick estimates of shaking intensities from earthquakes, including estimates of population impacted by the various intensity levels.

Gap Analysis Program
http://gapanalysis.nbii.gov/portal/server.pt?open=512&objID=1482&mode=2&in_hi_userid=2&cached=true.

Ecological data, consisting mostly of land use/land-cover maps and ecological zones.

Software

EQECAT Inc. Catastrophe Models

Described in Eguchi, et al. (1997), they sell software for the estimation of losses due to natural disasters.

FEMA,
HAZUS-MH
http://www.fema.gov/plan/prevent/hazus/index.shtm.

Hazards U.S. (HAZUS) is a software tool designed to provide individuals, businesses, and communities with information and tools to mitigate natural hazards. HAZUS is a natural hazard loss estimation software program developed by the National Institute of Building Sciences (NIBS) with funding from FEMA. It allows users to compute estimates of damage and losses from natural hazards using geographical information systems (GIS) technology. Originally designed to address earthquake hazards, HAZUS has been expanded into HAZUS Multi-Hazard (HAZUS-MH), a multi-hazard methodology with new modules for estimating potential losses from wind (including hurricane) and flood hazards. NIBS maintains committees of wind, flood, earthquake, and software experts to provide technical oversight and guidance to HAZUS-MH development. HAZUS-MH uses ArcGIS software to map and display hazard data and the results of damage and economic loss estimates for buildings and infrastructure. The ArcGIS software, developed by ESRI, is required to run HAZUS and is sold separately. Three data input tools have been developed to support data collection: the Inventory Collection Tool, Building Inventory Tool, and Flood Information Tool. The Inventory Collection Tool helps users collect and manage local building data for more refined analyses than are possible with the national level data sets that come with HAZUS. The Building Inventory Tool allows users to import building data from large datasets, such as tax assessor records. The Flood Information Tool helps users manipulate flood data into the format required by the HAZUS flood model. FEMA has also developed a companion software tool called the HAZUS-MH Risk Assessment Tool to produce risk assessment outputs for earthquakes, floods, and hurricanes. The Risk Assessment Tool pulls natural hazard data, inventory data, and loss estimate data into pre-formatted summary tables and text. HAZUS-MH can provide a probability of minor damage, moderate damage, and the destruction of constructed facilities. This information is viewable by occupancy type, type of building, essential facilities, and by user defined facilities.

GAMS
http://www.gams.com/.

A software package for creating and solving Computable General Equilibrium (CGE) models.

GEMPACK
http://www.monash.edu.au/policy/gempack.htm.

Another software package for solving CGE models.

IMPLAN
http://implan.com/v3/.

They are publishers of software, regional Social Accounting Matrices, and consulting services. Their products and services are geared toward regional economic modeling, and the estimation of the economic impacts of government policies and other shocks.

NIST
Cost-Effectiveness Tool for Capital Asset Protection, Version 4
http://www.nist.gov/bfrl/economics/CETSoftware.cfm

This paper describes the CET 4.0 software tool. It is intended to implement the ASTM E 2506 protocol for evaluating the cost effectiveness of proposed disaster mitigation measures. It takes as inputs costs of the various proposed alternatives, including O&M costs; costs associated with relevant disaster events, and the probabilities associated with the events. It is capable of handling uncertainty in costs of likelihood of adverse events using Monte Carlo methods.

Risk Management Solutions, Inc.,
http://www.rms.com/.

They provide data, software, and consulting services primarily to the insurance industry. Their software is designed to provide real-time estimates of losses due to natural disasters, both predicted and actual.

Appendix C Mathematical Appendix

C.1 What is the Optimal Recovery Path from Disaster[20]

Assume a single representative consumer who maximizes utility as:

$$\max \int_0^\infty \quad \text{---} \quad (1)$$

Subject to the budget constraint:

$$(2)$$

Where:

- C is a function representing consumption at any given time.
- K is a function representing amount of capital at any given time.
- L Labor, which is normalized to 1 for this problem.
- α Cobb-Douglass production parameter; related to capital intensity.
- δ depreciation rate of capital.
- ρ discount rate.
- σ preferences parameter; related to the intertemporal elasticity of substitution.
- x exogenous economic growth rate.

The Hamiltonian for the problem is:

$$\text{---} \quad (3)$$

The solution requires the following conditions be met:

$$\text{---} = 0$$

and

$$\text{---}$$

---

[20] This discussion is based on Barro and Sala-i-Martin (2003).

The following change of variables effectively normalizes for economic growth:

and

Which imply:

and

So, after the change of variables, our solution must meet the following three equations:

$$\qquad (4)$$

$$\qquad (5)$$

$$- \qquad (6)$$

Taking the derivative of the log of equation 5 gives:

$$-\log \quad - \quad \log$$

or

$$- \qquad -$$

Substituting into equation 6 gives:

$$- \qquad (7)$$

On the balanced-growth path both and are zero. Solving for c and k then gives:

The optimal recovery path was solved using numerical methods to jointly solve equations 4 and 7, and using the standard[21] parameter values of $\alpha = 0.75$, $\delta = 0.05$, $\rho = 0.02$, $\sigma = 3$, and $x = 0.02$.

---

[21] According to Barro and Sala-i-Martin (2003).

C.2 A Rationalization of the Multiplicative Resilience Index

Note that throughout this appendix we will suppress expression of t.

Let I be the set of individuals in a community.

Let J be the set of services provided.

Let J^* be the set of all subsets of J.

Let $q_{ij} \in [0,1]$ be a quality of service variable such that q_{ij} is the normalized quality of service individual i receives for service j.

The expression $|\cdot|$ returns the number of elements in the set.

Then the Quality of Service indicator for Service j is

$$Q_j = \frac{\sum_{i \in I} q_{ij}}{|I|}$$

And the MCEER combined quality of service is

$$\frac{\prod_{j \in J} Q_j}{\sum_{i \in I} \frac{1}{|I|}\left(1 - \prod_{j \in J} q_{ij}\right)}$$

As discussed in the text, this is a social welfare function, and there is no particular reason to think it is the only one.

To simplify our notation, we define the set function:

$$P(j, x) = \{i \in I : q_{ij} \geq x\}$$

That is, $P(j,x)$ is the set of all the people who have a quality of service of at least x for service j.

And, for some $S \subseteq J$, and vector $x \in [0,1]^{|S|}$

$$P(S, x) = \bigcap_{j \in S} P(j, x_j)$$

In this notation, it can be readily shown that:

$$Q_j = \frac{\sum_{i \in I} q_{ij}}{|I|} = \frac{1}{|I|} \int_0^1 |P(j, x)| \, dx$$

The main result is this theorem:

Theorem:

Given the following assumptions:

- Independence of service outages:

$$P\left(\bigcap_j A_j\right) = \prod_j P(A_j) \quad \text{for all } j.$$

- Leontief Utility:

    Each individual i, has the utility function $u_i = \min_j x_{ij}$.

- Additive welfare function U:

$$U = \frac{\sum_i u_i}{|I|}.$$

Then the aggregate welfare function is:

Proof:

$$U = \frac{\sum_i u_i}{|I|} = \frac{1}{|I|} \sum_i P(\bigcap_j A_{ij})$$

By independence we can rewrite this as:

$$\frac{1}{|I|} \sum_i \prod_j \frac{\sum_i P(A_{ij})}{|I|} = \prod_j \frac{\sum_i P(A_{ij})}{|I|}$$

A second application of independence gives us:

$$\prod_j \frac{\sum_i P(A_{ij})}{|I|} \quad \blacksquare$$

Aside from the implicit normalization, this is equivalent to the MCEER formulation.

The MCEER formulation is additionally normalized as:

$$\frac{\prod_j p_j}{\sum_j (1 - p_j) + \prod_j p_j}$$

102

If all services are either on or off then the $\in \{0,1\}$. Maintaining the assumption of independence, we can rewrite the MCEER formulation as:

$$\frac{\sum : \prod (1-)}{\sum : \sum (1-)}$$

In words, it is the ratio of people who have suffered no service outage to the number of people with some service.

References

Adams, Beverley, Charles Huyck, Babak Mansouri, Ronald Eguchi, and Masanobu Shinozuka. 2004. "Application of High-Resolution Optical Satellite Imagery for Post-Earthquake Damage Assessment: The 2003 Boumerdes (Algeria) and Bam (Iran) Earthquakes." *Research Progress and Accomplishments 2003-2004* 173-186.

Amendola, A. et al. 2000. "A Systems Approach to Modeling Catastrophic Risk and Insurability." *Natural Hazards* 21:381-393.

American Society of Civil Engineers. 2009. *2009 report card for America's infrastructure.* Reston Va.: American Society of Civil Engineers.

ASTM. 2006. "Standard Guide for Developing a Cost-Effective Risk Mitigation Plan for New and Existing Constructed Facilities."

Auf der Heide, Erik. 2006. "The Importance of Evidence-Based Disaster Planning." *Annals of Emergency Medicine* 47:34-49.

Barro, Robert J., and Xavier Sala-i-Martin. 2003. *Economic Growth, 2nd Edition.* 2nd ed. The MIT Press.

Belli, Anne, and Lisa Falkenberg. 2005. "24 nursing home evacuees die in bus fire." *Houston Chronicle*, September 24 http://www.chron.com/disp/story.mpl/topfront/3367696.html (Accessed July 14, 2010).

Berke, Philip, Jack Kartez, and Dennis Wenger. 2008. "Recovery after Disaster: Achieving Sustainable Development, Mitigation and Equity." *Disasters* 17:93-109.

Bernknopf, Richard, Laura Dinitz, Sharyl Rabinovici, and Alexander Evans. 2001. "A Portfolio Approach to Evaluating Natural Hazard Mitigation Policies: An Application to Lateral-Spread Ground Failure in Coastal California." *International Geology Review* 43:424-440.

Bolin, Robert, and Lois Stanford. 1991. "Shelter, Housing and Recovery: A Comparison of U.S. Disasters." *Disasters* 15:24-34.

Borden, Kevin A, and Susan Cutter. 2008. "Spatial patterns of natural hazards mortality in the United States." *International Journal of Health Geographics* 7:64.

Borden, Kevin A, M. C Schmidtlein, C. T Emrich, W. W Piegorsch, and Susan Cutter. 2007. "Vulnerability of U.S. cities to environmental hazards." *Journal of Homeland Security and Emergency Management* 4:1–21.

Bowden, Martyn, David Pijawka, Gary Roboff, Kenneth Gelman, and Daniel Amaral. 1977. "Reestablishing Homes and Jobs: Cities." Pp. 69-145 in Haas, J, Robert Kates, and Martyn Bowden, eds. *Reconstruction following disaster*. Cambridge Mass.: MIT Press.

Braun, Michael, and Alexander Muermann. 2004. "The impact of regret on the demand for insurance." *The Journal of Risk and Insurance* 71:737-767.

Brody, S. D, S. Zahran, W. E Highfield, H. Grover, and A. Vedlitz. 2008. "Identifying the impact of the built environment on flood damage in Texas." *Disasters* 32:1–18.

Bruneau, Michel et al. 2003. "A Framework to Quantitatively Assess and Enhance the Seismic Resilience of Communities." *Earthquake Spectra* 19:733-752.

Burby, Raymond, ed. 1998. *Cooperating with nature: Confronting natural hazards with land-use planning for sustainable communities*. Washington DC: Joseph Henry Press.

Carson, R, and W Hanemann. 2005. "Contingent Valuation." Pp. 821-936 in K.-G. Mäler and J.R. Vincent, Eds, *Handbook of Environmental Economics*, vol. 2. Elsevier.

Chandler, Tertius. 1987. *Four thousand years of urban growth: an historical census*. Lewiston N.Y. U.S.A.: St. David's University Press.

Chang, Semoon. 1983. "Disasters and Fiscal Policy: Hurricane Impact on Municipal Revenue." *Urban Affairs Review* 18:511-523.

Chang, Stephanie. 2010. "Urban disaster recovery: a measurement framework and its application to the 1995 Kobe earthquake." *Disasters* 34:303-327.

Chang, Stephanie, and Anthony Falit-Baiamonte. 2002. "Disaster vulnerability of businesses in the 2001 Nisqually earthquake." *Global Environmental Change Part B: Environmental Hazards* 4:59-71.

Chapman, Robert, and Amy Rushing. 2008. *Users Manual for Version 4.0 of the Cost-Effectiveness Tool for Capital Asset Protection*. Gaithersburg, MD: NIST.

Collins, Douglas, and Stephen Lowe. 2001. "A Macro Validation Dataset for U.S. Hurricane Models." in *Casualty Actuarial Society Forum*. Arlington, VA: Casualty Actuarial Society http://www.casact.org/dare/index.cfm?fuseaction=view&abstrID=3306 (Accessed March 16, 2010).

Comfort, Louise. 1999. *Shared Risk: Complex Systems in Seismic Response*. Pergamon.

Comfort, Louise, and Thomas Haase. 2006. "Communication, Coherence, and Collective Action: The Impact of Hurricane Katrina on Communications Infrastructure." *Public Works Management & Policy* 10:328-343.

Cutter, Susan. 2008. "A Framework for Measuring Coastal Hazard Resilience in New Jersey Communities." Working Paper.

Cutter, Susan, Bryan Boruff, and Lynn Shirley. 2003. "Social Vulnerability to Environmental Hazards." *Social Science Quarterly* 84:242-261.

Cutter, Susan, and C Emrich. 2005. "Are natural hazards and disaster losses in the U.S. increasing?." *EOS Transactions* 86.

Cutter, Susan, Melanie Gall, and Christopher Emrich. 2008. "Toward a comprehensive loss inventory of weather and climate hazards." Pp. 279-295 in Diaz, Henry, and Richard Murnane, eds. *Climate extremes and society*. Cambridge University Press.

Dahlhamer, James, and Kathleen Tierney. 1998. "Rebounding from Disruptive Events: Business Recovery Following the Northridge Earthquake." *Sociological Spectrum* 18:121-141.

Dlugolecki, Andrew. 2008. "An overview of the impact of climate change on the insurance industry." Pp. 248-278 in Diaz, Henry, and Richard Murnane, eds. *Climate extremes and society*. Cambridge University Press.

Dodo, Atsuhiro, Rachel Davidson, Ningxiong Xu, and Linda Nozick. 2007. "Application of regional earthquake mitigation optimization." *Computers & Operations Research* 34:2478-2494.

Dodo, Atsuhiro, Ningxiong Xu, Rachel Davidson, and Linda Nozick. 2005. "Optimizing Regional Earthquake Mitigation Investment Strategies." *Earthquake Spectra* 21:305-327.

Downton, Mary, J Zoe Barnard Miller, and Roger Pielke. 2005. "Reanalysis of U.S. National Weather Service Flood Loss Database." *Natural Hazards Review* 6:13-22.

Downton, Mary, and Roger Pielke. 2001. "Discretion without Accountability: Politics, Flood Damage, and Climate." *Natural Hazards Review* 2:157-166.

Downton, Mary, and Roger Pielke. 2005. "How Accurate are Disaster Loss Data? The Case of U.S. Flood Damage." *Natural Hazards* 35:211-228.

Ellson, Richard W., Jerome W. Milliman, and R. Blaine Roberts. 1984. "Measuring the Regional Economic Effects of Earthquakes and Earthquake Predictions." *Journal of Regional Science* 24:559-579.

Ermoliev, Yuri, Tatiana Ermolieva, Gordon MacDonald, Vladimir Norkin, and Aniello Amendola. 2000. "A system approach to management of catastrophic risks." *European Journal of Operations Research* 122:452-460.

FEMA. 2005. *Risk Assessment: A How-To Guide to Mitigate Potential Terrorist Attacks Against Buildings*. Washington DC: FEMA.

Foster, Kenneth, and Robert Giegengack. 2006. "Planning for a City on the Brink." Pp. 109-128 in Daniels, Ronald, Donald Kettl, and Howard Kunreuther, eds. *On risk and disaster: lessons from Hurricane Katrina*. Philadelphia: University of Pennsylvania Press.

Fothergill, Alice. 1998. "The Neglect of Gender in Disaster Work: An overview of the literature." Pp. 11-26 in Enarson, Elaine, and Betty Morrow, eds. *The gendered terrain of disaster: through women's eyes*. Westport Conn.: Praeger.

Friesma, H, J Caporaso, G Goldstein, R Linberry, and R McCleary. 1979. *Aftermath: Communities after natural disasters*. Beverly Hills, Calif: Sage.

Fritz, Charles. 1961. "Disaster." Pp. 651-694 in Merton, Robert, and Robert Nisbet, eds. *Contemporary social problems*. New York: Harcourt Brace World.

Gall, Melanie, Kevin A Borden, and Susan Cutter. 2009. "When Do Losses Count?." *Bulletin of the American Meteorological Society* 90:799-809.

Garrett, Thomas, and Russell Sobel. 2003. "The Political Economy of FEMA Disaster Payments." *Economic Enquiry* 41:496-509.

Glaeser, E. L, and J. D Gottlieb. 2009. "The wealth of cities: agglomeration economies and spatial equilibrium in the United States." *Journal of Economic Literature* 47:983–1028.

Godschalk, David, T Beatley, Philip Berke, D Brower, and E Kaiser. 1999. *Natural Hazard mitigation: recasting disaster policy and planning*. Washington DC: Island Press.

Gordon, P., H. W Richardson, and B. Davis. 1998. "Transport-related impacts of the Northridge earthquake." *Journal of Transportation and Statistics* 1:21–36.

Gordon, Peter, James Moore, and Harry Richardson. 2002. *Economic-Engineering Integrated Models for Earthquakes: Socioeconomic Impacts*. Pacific Earthquake Engineering Research Center: College of Engineering: University of California, Berkeley.

Grossi, Patricia, Howard Kunreuther, and Chandu Patel. 2005. *Catastrophe modeling: a new approach to managing risk*. Springer.

Guimaraes, Paulo, Frank Hefner, and Douglas Woodward. 1993. "Wealth and Income Effects of Natural Disasters: An Econometric Analysis of Hurricane Hugo." *Review of Regional Studies* 23:97-114.

H. John Heinz III Center for Science, Economics, and the Environment. 2000. *The hidden costs of coastal hazards: implications for risk assessment and mitigation*. Washington D.C.: Island Press.

Haas, J, Robert Kates, and Martyn Bowden, eds. 1977. *Reconstruction following disaster*. Cambridge Mass.: MIT Press.

Haas, J, Patricia Trainer, Martyn Bowden, and Robert Bolin. 1977. "Reconstruction Issues in Perspective." Pp. 25-68 in Haas, J, Robert Kates, and Martyn Bowden, eds. *Reconstruction following disaster*. Cambridge Mass.: MIT Press.

Hallegatte, Stéphane. 2008. "An Adaptive Regional Input-Output Model and its Application to the Assessment of the Economic Cost of Katrina." *Risk Analysis* 28:779-799.

Harrald, John. 2006. "Agility and Discipline: Critical Success Factors for Disaster Response." *The Annals of the American Academy of Political and Social Science* 604:256-272.

Howe, Charles, and Harold Cochrane. 1993. *Guidelines for the Uniform Definition, Identification, and Measurement of Economic Damages from Natural Hazard Events*. Institute of Behavioral Science, University of Colorado.

Isumi, Masanori, Noriaki Nomura, and Takao Shibuya. 1985. "Simulation of Post-Earthquake Restoration for Lifeline Systems." *International Journal of Mass Emergencies and Disasters* 3:87-106.

Kappos, Andreas, and E. Dimitrakopoulos. 2008. "Feasibility of pre-earthquake strengthening of buildings based on cost-benefit and life-cycle cost analysis, with the aid of fragility curves." *Natural Hazards* 45:33-54.

Kates, Robert, and David Pijawka. 1977. "From Rubble to Monument: the Pace of Reconstruction." Pp. 1-24 in Haas, J, Robert Kates, and Martyn Bowden, eds. *Reconstruction following disaster*. Cambridge Mass.: MIT Press.

Kozin, Frank, and Huakang Zhau. 1990. "System Study of Urban Response and Reconstruction due to Earthquake." *journal of Engineering Mechanics* 116:1959-1972.

Kunreuther, H., R. Meyer, and C. Van den Bulte. 2004. *Risk Analysis for Extreme Events: Economic Incentives for Reducing Future Losses*. NIST.

Kunreuther, Howard. 2001. "Incentives of Mitigation Investment and More Effective Risk Management: the Need for Public-Private Partnerships." *Journal of Hazardous Materials* 86:171-185.

Kunreuther, Howard, and Erwann Michel-Kerjan. 2009. *At War with the Weather: Managing Large-Scale Risks in a New Era of Catastrophes*. The MIT Press.

Landsea, Christopher et al. 2003. *The Atlantic Hurricane Database Re-analysis Project Documentation for 1851-1910 Alterations and Addition to the HURDAT Database*. NOAA http://www.aoml.noaa.gov/hrd/hurdat/Documentation.html (Accessed March 16, 2010).

Liu, Haibin, Rachel Davidson, and T. Apanasovich. 2007. "Statistical Forecasting of Electric Power Restoration Times in Hurricanes and Ice Storms." *Power Systems, IEEE Transactions on* 22:2270-2279.

Loomes, Graham, and Robert Sugden. 1982. "Regret Theory: An Alternative Theory of Rational Choice Under Uncertainty." *The Economic Journal* 92:805-824.

Lutter, Randall, John Morrall, and W. Kip Viscusi. 1999. "The cost-per-life-saved cutoff for safety-enhancing regulations." *Economic Inquiry* 37:599-608.

Manyena, Siambabala. 2006. "The concept of resilience revisited." *Disasters* 30:433-450.

MCEER. 2010. "PEOPLES: A framework for defining and measuring disaster resilience." Working Paper.

Meacham, Brian, and Matthew Johann. 2006. *Extreme Event Mitigation in Buildings*. Quincy, Mass: National Fire Protection Association.

Miles, Scott, and Stephanie Chang. 2003. *Urban disaster recovery: a framework and simulation model*. MCEER.

Mileti, Dennis. 1999. *Disasters by Design: a reassessment of Natural Hazards in the United States*. Washington D.C.: Joseph Henry Press.

Mileti, Dennis, and J Sorensen. 1990. *Communication of emergency public warnings: a social science perspective and state-of-the-art assessment*. Oak Ridge National Laboratory.

Miller, Stuart, Robert Muir-Wood, and Auguste Boissonnade. 2008. "An exploration of trends in normalized weather-related catastrophe losses." Pp. 225-247 in Diaz, Henry, and Richard Murnane, eds. *Climate extremes and society*. Cambridge University Press.

Muermann, Alexander, and Howard Kunreuther. 2008. "Self-protection and insurance with interdependencies." *Journal of Risk and Uncertainty* 36:103-123.

National Fire Protection Association. 2007. *Standard on Disaster/Emergency Management and Business Continuity Programs: 2007 Edition*.

National Research Council. 1999. *The Impacts of Natural Disasters: A Framework for Loss Estimation*. Washington DC: National Academies Press.

National Science Technology Council. 2005. *Grand Challenges for Disaster Reduction*. Washington DC: Executive Office of the President.

NEHRP. 2010. "Comments on the Meaning of Resilience." http://www.nehrp.gov/pdf/ACEHRCommentsJan2010.pdf (Accessed May 5, 2010).

Norris, Fran, Susan Stevens, Betty Pfefferbaum, Karen Wyche, and Rose Pfefferbaum. 2008. "Community Resilience as a Metaphor, Theory, Set of Capacities, and Strategy for Disaster Readiness." *American Journal of Community Psychology* 41:127-150.

Perry, Ronald, and Michael Lindell. 2007. "Disaster Response." Pp. 159-181 in Waugh, William, and Kathleen Tierney, eds. *Emergency management: principles and practice for local government*. 2nd ed. Washington D.C.: ICMA Press.

Pielke, Roger, and Mary Downton. 2000. "Precipitation and Damaging Floods: Trends in the United States, 1932–97." *Journal of Climate* 13:3625-3637.

Pielke, Roger, and Christopher Landsea. 1998. "Normalized Hurricane Damages in the United States: 1925–95." *Weather and Forecasting* 13:621-631.

Quarantelli, E. 1983. *Delivery of emergency medical services in disasters: assumptions and realities*. New York N.Y.: Irvington Publishers.

Quarantelli, E. 1989. *Disaster Recovery: Comments on the Literature and a Mostly Annotated Bibliography*. Working Paper.

Quarantelli, E. 1982. *Sheltering and housing after major community disasters: case studies and general observations*. Disaster Research Center: University of Delaware.

RMS Inc. 2010. "RMS FAQ: 2010 Haiti Earthquake and Caribbean Earthquake Risk."

Rose, Adam. 2004. "Economic Principles, Issues, and Research Priorities in Hazard Loss Estimation." Pp. 13-36 in Okuyama, Yasuhide, and Stephanie Chang, eds. *Modeling Spatial and Economic Impacts of Disasters*. Springer.

Rose, Adam. 2002. "Model Validation in Estimating Higher-Order Economic Losses from Natural Hazards." Pp. 105-131 in *Acceptable Risk Processes: Lifelines and Natural Hazards*. Reston, Virginia: American Society of Civil Engineers.

Rose, Adam, Juan Benavides, Stephanie Chang, Philip Szczesniak, and Dongsoon Lim. 1997. "The Regional Economic Impact of an Earthquake: Direct and Indirect Effects of Electricity Lifeline Disruptions." *Journal of Regional Science* 37:437-458.

Rose, Adam, and Dongsoon Lim. 2002. "Business interruption losses from natural hazards: conceptual and methodological issues in the case of the Northridge earthquake." *Global Environmental Change Part B: Environmental Hazards* 4:1-14.

Rose, Adam, Gbadebo Oladosu, and Shu-Yi Liao. 2007. "Business Interruption Impacts of a Terrorist Attack on the Electric Power System of Los Angeles: Customer Resilience to a Total Blackout." *Risk Analysis* 27:513-531.

Rubin, Claire. 1985. *Community Recovery from a Major Natural Disaster*. Institute of Behavioral Science, University of Colorado.

Rubin, Claire B. 1991. *Disaster recovery after Hurricane Hugo in South Carolina*. Natural Hazards Research and Applications Information Center, Institute of Behavioral Science, University of Colorado.

SPUR. 2009. "The Resilient City." *Urbanist* 4-21.

Swiss Re. 2010. *Natural catastrophes and man-made disasters in 2009: catastrophes claim fewer victims, insured losses fall*. Swiss Re.

Thomas, Douglas, and Robert Chapman. 2008. *A Guide to Printed and Electronic Resources for Developing a Cost-Effective Risk Mitigation Plan for New and Existing Constructed Facilities*. Gaithersburg, MD: NIST.

Tierney, Kathleen. 1997a. "Business impacts of the Northridge earthquake." *Journal of Contingencies and Crisis Management* 5:87-97.

Tierney, Kathleen. 1997b. "Impacts of Recent Disasters on Businesses: The 1993 Midwest Floods and the 1994 Northridge Earthquake." Pp. 189-222 in *Economic consequences of earthquakes: Preparing for the unexpected*. NCEER.

Tierney, Kathleen. 2006. "Social Inequality, Hazards, and Disasters." Pp. 109-128 in *On risk and disaster : lessons from Hurricane Katrina*. Philadelphia: University of Pennsylvania Press.

Tierney, Kathleen, Michael Lindell, and Ronald Perry. 2001. *Facing the unexpected: disaster preparedness and response in the United States*. Washington D.C.: Joseph Henry Press.

U.S. Government Accountability Office. 2007. *NATURAL DISASTERS: Public Policy Options for Changing the Federal Role in Natural Catastrophe Insurance*. Washington DC: U.S. GAO.

Vigdor, J. 2008. "The economic aftermath of Hurricane Katrina." *The Journal of Economic Perspectives* 22:135–154.

Viscusi, W. Kip. 2009. "Valuing risks of death from terrorism and natural disasters." *Journal of Risk and Uncertainty* 38:191-213.

Vranes, K., and R. Pielke Jr. 2009. "Normalized Earthquake Damage and Fatalities in the United States: 1900–2005." *Natural Hazards Review* 10:84.

Waugh, William, and Kathleen Tierney, eds. 2007. *Emergency management: principles and practice for local government*. 2nd ed. Washington D.C.: ICMA Press.

Webb, Gary, Kathleen Tierney, and James Dahlhamer. 2000. "Businesses and Disasters: Empirical Patterns and Unanswered Questions." *Natural Hazards Review* 1:83-90.

Wenger, Dennis, E. Quarantelli, and R. Dynes. 1986. *Disaster Analysis: Emergency Management Offices and Arrangements*. Disaster Research Center: University of Delaware.

Wenger, Dennis, E. Quarantelli, and R. Dynes. 1989. *Disaster analysis: Police and fire departments*. Disaster Research Center: University of Delaware.

West, Carol, and David Lenze. 1994. "Modeling the Regional Impact of Natural Disaster and Recovery: A General Framework and an Application to Hurricane Andrew." *International Regional Science Review* 17:121-150.

White, Gilbert, Robert Kates, and Ian Burton. 2001. "Knowing better and losing even more: the use of knowledge in hazards management." *Global Environmental Change Part B: Environmental Hazards* 3:81-92.

Wright, J, P Rossi, S Wright, and E Weber-Burdin. 1979. *After the clean-up: Long-range effects of natural disasters*. Beverly Hills, Calif: Sage.

Xu, N., Rachel Davidson, L. K Nozick, and A. Dodo. 2007. "The risk-return tradeoff in optimizing regional earthquake mitigation investment." *Structure and Infrastructure Engineering* 3:133–146.

www.ingramcontent.com/pod-product-compliance
Lightning Source LLC
Chambersburg PA
CBHW081727170526
45167CB00009B/3735